mathematics for human flourishing

mathematics for human flourishing

Francis Su

WITH REFLECTIONS BY

Christopher Jackson

Yale

UNIVERSITY PRESS

New Haven and London

Yale University Press books may be purchased in quantity for educational, business, or promotional use. For information, please e-mail sales.press@yale.edu (U.S. office) or sales@yaleup.co.uk (U.K. office).

Designed by Nancy Ovedovitz and set in Minion and Scala Sans types by Tseng Information Systems, Inc. Chapter-opening art by Carl Olsen (carlolsen.net). Selected illustrations courtesy Barbara Schoeberl, Animated Earth LLC. Printed in the United States of America.

Library of Congress Control Number: 2019943551
ISBN 978-0-300-23713-9 (hardcover : alk. paper)
ISBN 978-0-300-25851-6 (paperback)

A catalogue record for this book is available from the British Library.

10 9 8 7 6 5 4

For the Christophers and Simones of the world

contents

preface

This book is not about how great mathematics is, though it is, indeed, a glorious endeavor. Nor does it focus on what math can do, though it undeniably can do many things. Rather, this is a book that grounds mathematics in what it means to be a human being and to live a more fully human life.

This book grew out of a speech I gave in January 2017 at the end of my term as the president of the Mathematical Association of America. Although I was addressing a conference of mathematicians, the underlying themes were universal and the message resonated in ways I could not have anticipated. The tearful response of the audience showed me that there truly is a need, even among those who do math for a living, to talk about our longings for the common good, and the need for us to be better human beings to one another. After the speech was reported in *Quanta Magazine* and *Wired,* I received numerous letters from people whose experiences of math matched my own: hurtful ones when it is not practiced well, and joyful ones when we see how different it can be.

To welcome everyone to this conversation, I've aimed this book at a wide audience—especially those of you who don't see yourselves as "math people." Maybe the way for you to see yourself in mathematics is not for me to convince you that math is great or that math does lots of wonderful things, but for me to show you that math is intimately tied to being human. For then your deepest human desires reveal your mathematical nature—and you need only to awaken it.

I won't assume much background; I know we've all had different experiences in math, and it's totally fine to come to this book as you are. I'll refer to some mathematical ideas here and there and try to connect them to things you may know, in much the same way that you might have casual conversations about philosophy or music or sports. You may be reading this book on behalf of someone you know who is learning mathematics, and for that reason I will sometimes give advice to those who teach. No matter who you are, I hope you'll read this book as an invitation and think of the ideas as conversation starters—in the home, in the classroom, or among friends—for how to imagine mathematics in a new way.

mathematics for human flourishing

1
flourishing

Every being cries out silently to be read differently.
Simone Weil

Christopher Jackson is an inmate in a high-security federal prison. He's been in trouble with the law since he was fourteen. He didn't finish high school, he had an addiction to hard drugs, and at age nineteen he was involved in a string of armed robberies that landed him in prison with a thirty-two-year sentence.

By now, you've probably formed a mental image of who Christopher is, and you might be wondering why I'm opening

with his story. When you think about who does mathematics, would you think of Christopher?

Yet he wrote me a letter after seven years in prison. He said:

> I've always had a proclivity for mathematics, but being in a very early stage of youth and also living in some adverse circumstances, I never came to understand the true meaning and benefit of pursuing an education. . . .
>
> Over the last 3 years I have purchased and studied a multitude of books to give me a profound and concrete understanding of Algebra I, Algebra II, College Algebra, Geometry, Trigonometry, Calculus I and Calculus II.

When you think about who does mathematics, would you think of Christopher?

Every being cries out silently to be read differently.

Simone Weil (1909–1943) was a well-known French religious mystic and a widely revered philosopher. She is probably less well known as the younger sister of André Weil, one of history's most famous number theorists.

For Simone, to *read* someone means to interpret or make a judgment about them. She's saying, "Every being cries out silently to be judged differently." I wonder if Simone was crying out about herself. For she, too, loved and participated in mathematics, but she often compared herself unfavorably to her brother. In a letter to a mentor, she wrote:

> At fourteen I fell into one of those fits of bottomless despair that come with adolescence, and I seriously thought of dying because of the mediocrity of my natural faculties. The exceptional gifts of my brother, who had a childhood

Simone Weil, around 1937.
Photo courtesy of Sylvie Weil.

> and youth comparable to those of Pascal, brought my own inferiority home to me. I did not mind having no visible successes, but what did grieve me was the idea of being excluded from that transcendent kingdom to which only the truly great have access and wherein truth abides. I preferred to die rather than live without that truth.[1]

We know Simone loved mathematics because she used mathematical examples throughout her philosophical writing.[2] And you'll find her with André in photos of Bourbaki, a group of reformist French mathematicians, as a conspicuously lone woman. Their prank-filled meetings were perhaps not the most inviting places to be a woman.[3]

A meeting of Bourbaki, around 1938. Simone Weil is seen at left, leaning over her notes. André Weil is waving the bell. Photo courtesy of Sylvie Weil.

I often wonder what her relationship to mathematics would have been like if she hadn't always been in André's shadow.[4]

Every being cries out silently to be read differently.

I am a joyful enthusiast of mathematics, a teacher of mathematics, a researcher in mathematics, and former president of the Mathematical Association of America. So you might think that my relationship to mathematics has always been solid. I don't like the word *success,* but people think of me as successful, as if the true measure of my mathematical achievement were the awards I've received or the papers I've published. Even though

I've had advantages, including a middle-class background and parents who pushed me to excel, my pursuit of math, even for nobler reasons than achievement, has had its obstacles.

As a child I was attracted to the beautiful ideas of mathematics, and I longed to learn more. But I grew up in a small rural town in south Texas, with limited opportunities. There were few advanced math or science courses available in my high school, since college wasn't a standard option for its students. I didn't have a large network of friends excited about math. My parents, as motivated as they were to help me learn, didn't know where to look to nourish my interest in mathematics; finding such resources was even harder in the era before the internet existed. I mainly relied on older books from the public library. My love for math deepened when I was an undergraduate at the University of Texas, and I was admitted to Harvard for my PhD. But I felt out of place there, since my college degree was not from an Ivy League school and, unlike many of my peers, I did not have a full slate of graduate courses when I entered. I felt like Simone Weil, standing next to future André Weils, thinking I would never be able to flourish in mathematics if I was not like them.

I was told by one professor, *You don't have what it takes to be a successful mathematician.* That unkind remark forced me to consider, among other things, why I wanted to do mathematics. To do mathematics means more than just learning the facts of mathematics—it means seeing oneself as a capable mathematical learner who has the confidence and the habits of mind to tackle new problems. Unexpectedly, I joined the company of the multitude of people who have been wounded by harsh judgments and have questioned their capacity for math. There are many others who question the point of learning math, and still others who don't have access to a quality mathematical education. In the face of so many obstacles, it is a fair question for all of us to consider:

Why do mathematics?

Why was Christopher sitting in a prison cell studying calculus, even though he wouldn't be using it as a free man for another twenty-five years? What benefit does mathematics have for him? Why was Simone so captivated by transcendent mathematical truths? What do they offer, that she so desperately yearned to know more? Why should you persist in learning math or in seeing yourself as a mathematical explorer when others are telling you in subtle and not-so-subtle ways that you don't belong?

In this present moment, society is also asking what its relationship with mathematics should be. Is mathematics only a tool to make you "college- and career-ready" so you can achieve your real aims in life? Or is mathematics unnecessary for most of us and relevant only to an elite few? What value is there in studying math if you'll never use what you're learning? Tomorrow's jobs may not even use the math you learn today.

Amid the great societal shifts wrought by the digital revolution and the transition to an information economy, we are witnessing the rapid transformation of the ways we work and live. Mathematical tools are now prominent in every sector of the workforce, including the most dominant ones; presently, technology companies are the four most valuable companies in the world.[5] This means that power is now even more vested in those with mathematical skills.[6] In the span of a young person's lifetime, the tools of our daily lives have become mathematical as well. Search engines now satisfy our every investigative whim, with algorithms powered by linear algebra and advertising powered by game theory. Smartphones have become our digital butlers, storing our data in algebraically locked closets, interpreting our voice commands with statistical sensibilities, and pleasing us with a selection of analytically decompressed music.

Yet society has not taken seriously its obligation to provide a vibrant mathematics education for everyone. In many schools, teachers lack sufficient support. Outdated curricula and pedagogies prevent many students from experiencing math as a fascinating area of exploration, culturally relevant and important in all spheres of life. We hear voices in the public square saying that high school students don't need algebra, or that few people need to be good at math—implying that math is best left to the mathematicians.[7] Some college mathematics faculty effectively declare the same thing by abdicating the teaching of introductory classes, or by viewing the undergraduate math degree as only a pipeline for the production of math PhDs. Over many decades and at all levels—elementary school through college—there have been calls to change the way math is taught;[8] nevertheless, change has been slow, in part because the math curriculum has often served as a backdrop to political quarrels over the nature of education itself.[9]

We are not educating ourselves as well as we should, and like most injustices, this especially harms the most vulnerable. Lack of access to mathematics and lack of welcome in mathematics have had devastating consequences for the poor and other disadvantaged groups.[10] Not tapping everyone's potential is a loss for all of us and will limit the ability of future generations to solve the problems they will face.

Our failure to invest in people is already affecting us now. We are easily manipulated when we don't understand how new technologies work but expect them to make decisions on our behalf. We've been unaware of the ways that algorithms are used to sort us and track us and divide us—showing us different news, selling us different loans, and stirring different emotions in us than in our neighbors.[11] We witness entrepreneurs unwilling to critique the technologies they are inventing, politicians unable to hold them accountable because of a lack of mathematical sophistication, and a general public unprepared to contemplate its relationship to these technologies.

We all know there's math under the hood, but otherwise math seems cold, logical, and lifeless. No wonder we don't feel a personal connection to it. No wonder we don't feel a responsibility for how it's used.

You and I can change that. All of us have the capacity to embrace the wonder, power, and responsibility of mathematics by nourishing our affection for it. The need to do so in today's world cannot be overstated, and the stakes are high.

A society without mathematical affection is like a city without concerts, parks, or museums. To miss out on mathematics is to live without an opportunity to play with beautiful ideas and see the world in a new light. To grasp mathematical beauty is a unique and sublime experience that everyone should demand.

All of us—no matter who you are or where you're from—can cultivate mathematical affection. All of us can have a different relationship to mathematics than we imagined. All of us can read ourselves and one another differently.

I'm speaking to the demoralized, who've been injured by words someone said about their math abilities. I'm speaking to the disenchanted, for whom math has become boring. I'm speaking to those who haven't had the resources or the confidence to get a mathematical education but have always been curious about how things work. I'm speaking to the artist who never thought math was beautiful, the social worker who never thought math was relational, and the mathematician who never thought math was accessible to anyone else.

And I'm speaking to both those who teach math and those who think they'll never teach math—because *every single one of us is, whether we realize it or not, a teacher of math.* We all communicate attitudes about mathematics through what we say to others, and our words have indelible effects. You can communicate negativity: "I was never any good at math." "That subject is for boys." "Don't hang out with her—she's a nerd." "Son, I'm not a math person, and you probably have my genes." "Why would you take another math class?" Or you can communicate positivity: "Math is an exploratory adventure." "You *can* improve your math skills, just like I can improve my free throws." "Math is power to see hidden patterns." "Everyone has promise in mathematics."

You may be a parent someday, or an aunt or uncle, a youth group leader, a community volunteer, or in another position where you influence others—if so, you will be a teacher of math. If you help kids with their homework, you are a teacher of math. If you are afraid to help kids with their homework, you are teaching an attitude about math. Studies show that parents with math anxiety pass on that anxiety to their children. In fact, math-anxious parents are more likely to pass on math anxiety if they try to help their kids with their math homework than if they don't.[12] So your disposition toward math matters as much for a child's sake as for your own.

Reading ourselves differently will require all of us—those who've failed at math and those who've been successful—to change our view of what mathematics is and who should be learning it. This will require teachers to change their view of how they should be teaching it. We'll need to speak about it in a different way—and if we do, more of us will be drawn to mathematics when we see how mathematics connects to our deepest human desires.

So if you ask me, "Why do mathematics?" I will say this: "Mathematics helps people flourish."

Mathematics is for human flourishing.

Human flourishing refers to a wholeness—of being and doing, of realizing one's potential and helping others do the same, of acting with honor and treating others with dignity, of living with integrity even in challenging circumstances. It is not the same as happiness, and it is not just a state of mind. The well-lived life is a life of human flourishing. The ancient Greeks had a word for human flourishing—*eudaimonia*—which they viewed as the highest good: "the good composed of all goods; an ability which suffices for living well."[13] There is a similar word in Hebrew—*shalom*—which is used as a greeting. *Shalom* is sometimes translated as "peace," but the word has a far richer context. To say *shalom* to someone is to wish that they will flourish and live well. And Arabic has a related word: *salaam*.

A basic question, taken up by human beings throughout the ages, is: How do you achieve human flourishing? What is the well-lived life? The philosopher Aristotle said that flourishing comes through the exercise of virtue. The Greek concept of virtue is excellence of character that leads to excellence of conduct. So it includes more than just moral virtue; for instance, traits like courage and wisdom and patience are also virtues.

I claim that the proper practice of mathematics cultivates virtues that help people flourish. These virtues serve you well no matter what profession you choose or where your life takes you. And the movement toward virtue is aroused by basic human desires—the universal longings that we all have—which fundamentally motivate everything we do. These desires can be channeled into the pursuit of mathematics; the resulting virtues can enable you to flourish.

Consider this analogy: if doing mathematics is like navigating a sailboat, then human desires are the winds that power the sails, and virtues are the qualities of character that sailing builds—mindfulness, attention, and harmony with the wind. Of course, sailing is useful in getting us from point A to point B, but that is not the only reason to sail. And there are technical skills that must be mastered to sail well, but we don't learn sailing to become better at tying knots. Similarly, math skills are valuable, but they cannot serve as goals. The skills society needs from math may change, but the virtues needed from math will not.

In promoting the human side of mathematics, I join a growing chorus of those who have called for a humanization of mathematics and math education, often in service of addressing long-standing inequities, by shifting away from contextless portrayals of mathematics to reveal its social and cultural dimensions.[14] That laudable goal will not be possible—and will often be resisted—unless we name a purpose for learning mathematics that is more than just memorizing procedures for an eventual career.

When some people ask, "When am I ever going to use this?" what they are really asking is "When am I ever going to value this?"[15] They're equating math's value with utility because they haven't seen that they can value anything more. A grander, more purposeful vision of mathematics would tap into the desires that can entice us to do mathematics as well as the virtues that mathematics can build.

Therefore, each of the following chapters is devoted to one basic human desire whose fulfillment is a sign of human flourishing. In each, I illustrate how the pursuit of mathematics can meet this desire, and I illuminate the virtues that are cultivated by engaging in math in this way. Changing the practices of mathematics so that they do, in fact, meet these desires is our common responsibility if everyone is to flourish in mathematics.

I know some may hear me broach virtue and think I'm saying, Math makes you better than other people. No—I'm not saying that math gives you greater claim to worthiness or human dignity. I'm saying that the pursuit of math can, if grounded in human desires, build aspects of character and habits of mind that will allow you to live a more fully human life and experience the best of what life has to offer. None of us is wholly virtuous; we all are works in progress with room to grow. And there are many ways to grow in virtue, not just in mathematics. But does the proper practice of mathematics build *particular* virtues, like the ability to think clearly and to reason well? Unequivocally yes, and it may do so in a distinctive way.

Because I speak so highly of mathematics, you may think I idolize mathematics as the ultimate pursuit, to be prized above all other pursuits in life. That is not the case either—we must each discover what gives our soul its greatest purpose. Still, mathematics is a marvelous human endeavor—worth the effort to explore and participate in, and worth the effort to help others do the same—because it meets basic human desires and contributes in unique ways to a life lived well.

I hope that by seeing yourself in these desires, you can see yourself as a *mathematical explorer,* who can think in mathematical ways and who is welcome in mathematical spaces. And where the practice of mathematics is not yet grounded in these desires, I hope you will join me in changing it. In doing so, you will have new ways to experience mathematics, not just as a toolbox of facts and skills, but as a force for the flourishing of all.

Mathematical exploration begins with questions. So I'm scattering a few puzzles throughout this book. No pressure—you can skip them if you wish, or think about just the ones that seem enticing. Hints and solutions can be found in the back, but before you look there, I recommend playing around with each problem.

DIVIDING BROWNIES

A father bakes brownies in a rectangular pan as an after-school snack for his two daughters. Before his daughters get home, his wife comes along and removes a rectangle from somewhere in the middle, with the sides of the rectangle not necessarily parallel to the sides of the pan.

How can he make one straight cut and divide the remainder of the brownies evenly between his two daughters so that they get the same area?

A version of this puzzle was featured on the NPR show *Car Talk.*[a]

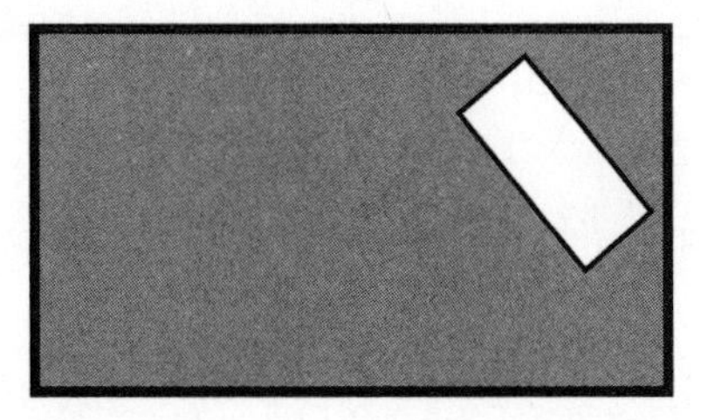

TOGGLING LIGHT SWITCHES

Imagine 100 lightbulbs, each with a switch numbered 1 through 100, all in a row, and all lights off. Suppose you do the following: toggle all switches that are multiples of 1, then toggle all switches that are multiples of 2, then toggle all switches that are multiples of 3, etc., all the way to multiples of 100. (To toggle a switch means to flip it on if it's off and off if it's on.)

When you are done, which lightbulbs are on and which are off? Do you see a pattern? Can you explain it?

a. See https://www.cartalk.com/puzzler/cutting-holey-brownies.

Nov. 26, 2013

Hello, Mr. Su. My name is Christopher Jackson, an inmate at United States Penitentiary McCreary in Pine Knot, Kentucky. I am 27 years old and I've been in prison a little over 7 years. I have a 32 year sentence for a string of armed robberies I committed in which no one was hurt, at the age of 19, while suffering from a severe addiction to a multitude of hard drugs.

I've always had a proclivity for mathematics, but being in a very early stage of youth and also living in some adverse circumstances, I never came to understand the true meaning and benefit of pursuing an education. At the age of 14, I became more involved with the juvenile justice system, because I left the house of my guardians, soon after I left off from attending school, and began becoming more involved in a criminal lifestyle. At the age of 17, at the prodding of my case manager and others supporting me, I got my G.E.D. and enrolled in Atlanta Technical College, but didn't attend but for a few days, and returned back deeper into my life of crime. Over the next couple of years, I struggled with addiction while going back and forth to jail until I committed the crimes I am now serving this sentence for. After pleading guilty to two charges in my indictment at the age of 21, I was remanded to federal custody and sent here to this facility, where I have spent the last 4 years of my life.

Over the last 7 years, I have developed an acute interest in studies and books concerning philosophy, mathematics, finance, economics, business, and politics. And over the last 3 years I have purchased and studied a multitude of books to give me a profound and concrete understanding of Algebra I, Algebra II, College Algebra, Geometry, Trigonometry, Calculus I and Calculus II.

The vast majority of problems in my life have stemmed from my hardheartedness and my unwillingness to listen to people who actually did know better than me or had positions of authority. Because even with a missing and deceased father, I had a mother, aunties, and a grandmother that really and actually attempted to raise me right. Every day that I wake up and go about my life I try not to allow the present circumstances that I've created for myself to determine the passions and the future I want to see myself in.

I became aware of your institution from a book I read by a Professor who teaches there, and by your institution being mentioned a couple of times on a television show I watch regularly.

I am a person of limited means, but I would like to know if there is a program that you have so I can by way of correspondence earn a degree in the art of mathematics from your school. I know you are a very busy person so I wouldn't want to take up any more of your time. But thank you for your time and consideration.

Christopher Jackson

2
exploration

It is like being lost in a jungle and trying to use all the knowledge that you can gather to come up with some new tricks, and with some luck you might find a way out.
Maryam Mirzakhani

The world of math is more weird and wonderful than some people want to tell you.
Eugenia Cheng

My friend Christopher Jackson is a mathematical explorer. He's limited by circumstance, but not by imagination. He's curi-

ous and inventive; he's fearless and persistent. Chris likes the challenge a good problem can bring.

For the past several years, Chris has been on a journey. He's exploring mathematics with fresh eyes, beginning to see that it's different from the dry and uninspired form of math he'd been taught before. He's growing in his knowledge and love of the subject, in spite of his isolation in prison and the difficulties that brings. I've had the privilege of observing, at a distance, his transformation.

Life hasn't been easy for Chris. He grew up in a working-class neighborhood of Augusta, Georgia, raised by his mother, with help from his aunts and his grandmother. Chris never knew his father, who succumbed to a crack cocaine addiction and died tragically in a car accident before Chris was two, hit by an eighteen-wheeler on a highway. Chris had some good influences—his mother instilled a love for books in him by reading to him frequently—but he had some negative influences as well. That led, in his teenage years, to drug addiction and the series of crimes that he described in his first letter to me.

I was cautious of, but also captivated by, that letter from the Pine Knot penitentiary in November 2013 (see facing page). It was handwritten, with orderly penmanship and earnest lines. I imagined a young man who had labored with care over every word. I couldn't see him—I could know him only through his writing—but perhaps the content of his character was more visible that way. I was touched by Chris's reflections about his troubled past, the future he would like to see himself in, and the way he had pursued his interest in math by teaching himself from books. I regretted that my school didn't have a distance program for him.

Chris and I have corresponded by mail, off and on, for six years now. We've talked about our mutual interest in mathemat-

proclivity for mathematics, but being in a very early stage of youth and also living in some adverse circumstances, I never came to understand the true meaning of and benefit of pursuing an education. At the age of 14, I began becoming more involved

ics and about life. With Chris's permission, I'm sharing excerpts from our correspondence, because his insights and experience amplify everything I am saying in this book. This is not a story about how I helped Chris do math in prison. Rather, this is a story about how Chris is coming to see himself, and mathematics, in a new way. His insights and his journey have inspired *me,* in writing a book about flourishing, to believe more fully that mathematics has something to offer everyone.

I am a mathematical explorer. My journey has been different than Chris's, but we've both been enticed by the power of mathematical exploration to awaken the imagination. As a child, I loved the stars, and in the rural Texas town where I lived, far from any big city, I could see even the dimmest ones. I begged my parents for a telescope, but we didn't have the money. So I devoured books on astronomy and dreamed about space. I wanted to be an astronaut, to visit other worlds and encounter strange new life-forms. That seemed exciting, until I realized how long it would take to travel to even the nearest star and thought about all the people I'd have to leave behind. But that didn't stop me from fantasizing. I fed myself a steady diet of science fiction, captivated by stories like Isaac Asimov's *Nightfall,* a novella about what happens when night finally comes to civilization on a planet with six suns. I could visit this world in my mind.

My childhood imagination was further stoked by the journeys of the *Pioneer* and *Voyager* probes through the solar sys-

Mimas by Saturnshine. The moon Mimas is being illuminated by sunlight reflected off Saturn. The Cassini division is the largest gap in the rings, visible on the left side of the photo. Image courtesy NASA/JPL-Caltech/Space Science Institute. Taken by the *Cassini* spacecraft on February 16, 2015.

tem in the late 1970s and early 1980s. For the first time, scientists captured up-close photographs of the moons of Jupiter and the rings of Saturn. Reaching these worlds required years of creative planning for all scenarios, good and bad, that these crewless probes might encounter. Just as the scientists themselves were making their discoveries from a distance, I could vicariously explore these worlds from a little town in south Texas. I loved poring over the *Voyager* images printed in the newspaper.

Mathematics can literally be seen in these worlds. The rings of Saturn encircle the planet in its equatorial plane. From a distance they look like stationary annular bands, but the rings are basically made of a multitude of boulder-size rocks (moonlets) that contain mostly ice and orbit the planet because of the force of gravity. The astronomer Galileo Galilei first observed the rings through a telescope in 1610. Unsure of what they were, he playfully referred to them as ears.[1] Later astronomers identified these structures as rings, with gaps between them. The *Voyager* probes enabled us to see the structure of the rings in finer detail, such as their pattern of high- and low-density ripples, much like the grooves on an old vinyl record.

Some of the structure of the rings, as I learned, can be explained with mathematical insight. All of the icy rocks at the same distance from Saturn take the same time to do one orbit—that's called their orbital period. The rocks that are farther away from Saturn have longer orbital periods and slower speeds than the ones that are closer, because they are less influenced by the planet's gravity. Think of the rings like a racetrack around Saturn, in which runners on the inner lanes go faster and also have less distance to travel than those on the outer lanes.

Something special happens when the orbital period of a rock is in one of certain exact whole-number ratios with the period of a moon of Saturn. For instance, suppose you have a rock and a moon revolving around Saturn such that in the time it takes the moon to do one orbit on an outer lane, the rock does precisely two on an inner lane. Every two orbits, the rock will pass the moon in exactly the same location in the rock's orbit.

The moon's strongest gravitational tug on the rock happens at these times of closest approach. Because these repeated tugs happen in the same location, they tend to reinforce each other and perturb the rock's orbit, much as pushing a child on a swing in sync with its motion causes the child to go higher. Thus, all the rocks at that distance from Saturn, with the same orbital period, tend to swing out of that orbit. This effect is called *resonance,* and when it is very strong it can create a gap in the rings.

The largest gap, some 3,000 miles wide, is called the Cassini

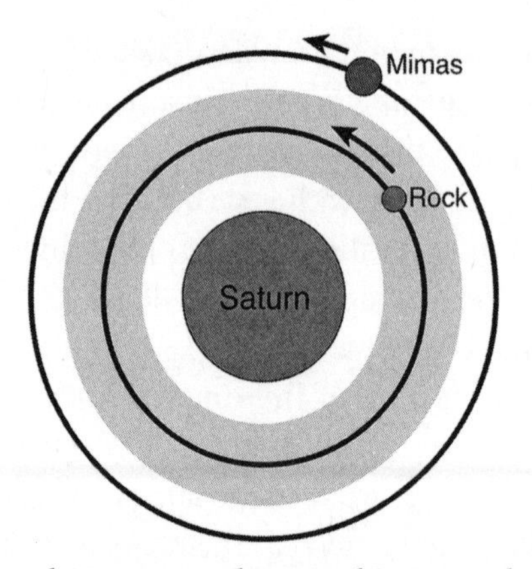

Icy rock in an inner orbit around Saturn catching up with Mimas. The effects of Mimas's gravity may accumulate and perturb the rock's orbit if the rock always passes the moon in the same location.

division and is the result of a two-to-one resonance between orbiting rocks and Saturn's moon Mimas. You can get resonance effects for other small-whole-number ratios of moon-to-rock periods (like three-to-two or four-to-three), though they are less pronounced, often more like ripples than gaps. Resonance effects between moons and rocks can explain many features in Saturn's rings.[2] In effect, we are seeing numerical ratios—simple fractions—whipping up visible patterns in the delicate orbital dances of these icy rocks! For a kid like me, it was enchanting that a little exploration, using mathematics and the imagination, could shed insight on objects 900 million miles away.

Mathematical exploration is very much like space exploration, but of a different kind of space—a space of ideas. You don't know what you'll find when you start. You send out probes to test theories. You are captivated by mystery, motivated by questions, undeterred by setbacks. You make discoveries from a distance: because the ideas themselves are not physical, you access this space through reason. Exploration and understanding are at the heart of what it means to do mathematics.

Unfortunately, *exploration* is not a word one associates with mathematics if one thinks it is just arithmetic—or something advanced and *even more* dreary that was discovered and settled long ago.

School mathematics sets you up for future exploration, but imagine how different our experiences would be if we could explore math *now*, as we learn it. Picture what it would be like to learn the rules of basketball and practice only free throws but never see a game and never play—until you're ready to go professional.[3] Learning wouldn't have been joyful, and you wouldn't now be prepared.

Exploration is a deep human desire and a mark of human flourishing. You don't need a lot of resources, except your mind, to be a math explorer, and as a result, you can embark on an adventure from anywhere—a prison, a small rural town, a far-flung corner of the earth. It is no surprise, then, that math explorers can be found in every society throughout history. This is most readily apparent in the games that people play, especially games of strategy, which generate interesting mathematical questions.

Achi is a game played by the Ashanti people of Ghana in West Africa. This two-player game is played on a three-by-three diagram of horizontal and vertical lines and two diagonals. Achi is like tic-tac-toe, but with a twist. The players each have just four pieces, which they take turns placing on the board's nine positions. The object is to get three in a row along one of the straight lines. If neither player has gotten three in a row by the time all

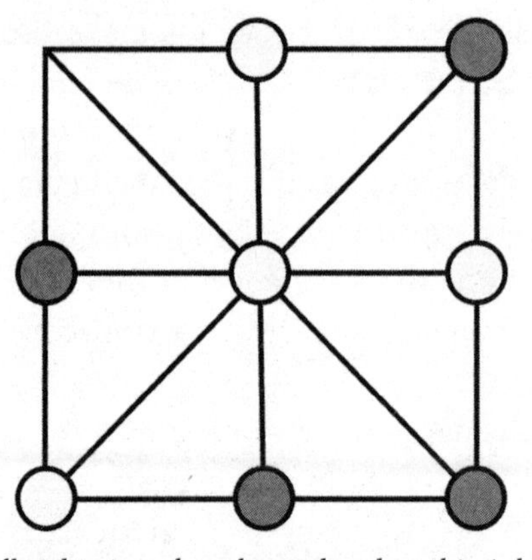

All eight pieces have been placed on this Achi board and there is no winner yet, so the players will take turns pushing one of their pieces onto the empty position until one gets three in a row.

the pieces have been placed, then there is a single empty space on the board. At that point the game moves into a second phase, in which the players take turns pushing one of their pieces along a line onto whatever space is empty. No jumping is allowed. The player who first gets three in a row wins.[4]

This is the standard description of Achi, but the rules leave some ambiguity.[5] For instance, in the second phase, what happens if a player is stuck, with no move to make? Is it impossible for a player to get stuck if both players play smartly (i.e., don't pass up any obvious opportunities to win)? You'll also need to decide whether players should be forced to make a move if they have moves to make. Mathematical reasoning can help you answer these questions and decide which options generate a more interesting game. In these variants, can the game of Achi go on forever, or can one player have a winning strategy (a plan of moves to guarantee a win no matter what the other player does)? What if each player has only three pieces instead of four? Can you create an interesting variant of Achi with different lines?

If you ask questions like these, you are a mathematical explorer. You are exploring the space of all possible ways that the game play could proceed. You send out "probes" by trying things. You don't know what you'll find when you start. When you discover answers using mathematical reasoning, you are doing it from a distance, because you can know things about how the game will play out *without* actually playing out every possible scenario.

For instance, in the game of tic-tac-toe there's a clever argument, called a *strategy-stealing argument,* to show that the second player cannot have a winning strategy: If she did have such a strategy, the first player could, by ignoring his own first move, pretend that he was the second player and use her strategy in responding to her! If that strategy ever suggested playing a move that he'd already made, he could just place his next move anywhere else—that's an extra move for him, and in tic-tac-toe any extra move can only help a player to win. The fact that both players can force a win is a contradiction and means that our assumption that the second player had a winning strategy was false. So the first player must be able to force a win or a draw. It's amazing that we can deduce this through mathematical reasoning—without playing a single game of tic-tac-toe.

Every culture plays games of strategy.[6] Every culture has math explorers, because strategic thinking is mathematical thinking. One of the best ways to claim your heritage in mathematics is to find a game of strategy from your own cultural history and embrace the kind of thinking the game requires. Probe it with exploratory questions.

Mathematical exploration begins with questions. The only requirement to be a math explorer is the ability to ask questions like *Why?* and *How?* and *What happens if . . . ?* All children do this, yet somewhere along the way, some stop asking questions—perhaps because they are told to memorize things and not to understand things. They are taught to follow procedures rather than to explore why those procedures work. They begin to think that there is just one right way to solve a problem rather than developing their own path to a solution. At every opportunity we need to counter the idea that math is memorization, and replace it with the idea that math is exploration. A math memorizer doesn't know how to react in unfamiliar situations, but a math explorer can flexibly adapt to changing conditions, because she has learned to ask questions that will prepare her for many scenarios. Effective math teachers know how to coax us to explore. The math teacher Fawn Nguyen has this advice for other teachers: "Critique the effectiveness of your lesson, not by what answers students give, but by what questions they ask."[7]

Exploration cultivates the virtue of *imagination.* In order to solve problems, you have to ideate new possibilities. To explain the orbital distances of the six planets known at the time, the German astronomer Johannes Kepler proposed a theory in his book *Mysterium Cosmographicum* (Cosmic Mystery) of six orbital spheres separated by (and tangent to) a succession of the five Platonic solids: the tetrahedron, cube, octahedron, icosahedron, and dodecahedron.

This theory didn't really fit the data, and we now know it is totally wrong, but it was imaginative! Brainstorming necessarily generates fancifully wrong ideas, but even wrong ideas soften the soil in which good ideas can grow. The same thing happens in math when you try to solve a hard problem. You'll get nowhere unless you start somewhere. Conversations among pro-

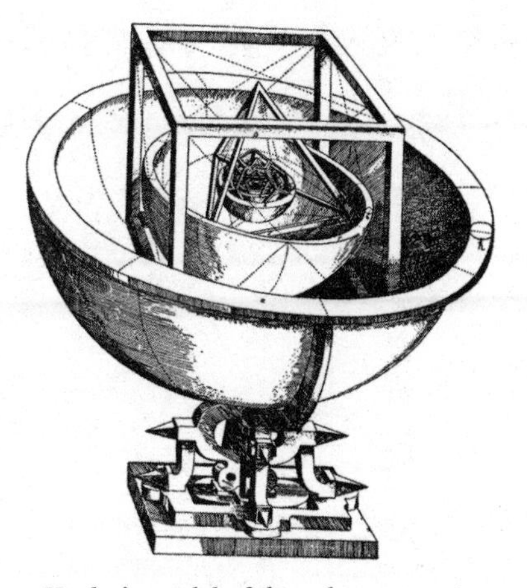

Kepler's model of the solar system in
Mysterium Cosmographicum.

fessional mathematicians often start like this: "Maybe we can show X or Y." Then they try the approach and realize it doesn't work, but the attempt itself can reveal new insights.

Exploration stimulates the virtue of *creativity.* The challenges of exploration often require new tools to solve problems that arise. For instance, the drive to reach the moon generated many inventions, now used in everyday items like cordless tools, memory foam, home insulation, and scratch-resistant lenses. Similarly, basic research in mathematics has often led, years later, to spectacular applications. The quest to understand primes has led to applications in cryptography; the topological theory of knots now has applications in protein folding; the theory of Radon transforms now animates the mathematics behind CAT scans.[8] Any interesting, well-designed math problem, even a simple one, will stretch your creativity. Good teachers know them, good puzzle books have them, math competitions curate them, and math explorers share them.[9] In his delightful

Create two rectangles so that the first has a bigger perimeter, and the second a bigger area.

A Bad Drawing, courtesy of Ben Orlin, from his book *Math with Bad Drawings.*

book *Math with Bad Drawings,* the math teacher Ben Orlin discusses the difference between a dull problem and an exploratory problem.[10] He gives this example:

> Find the area and perimeter of a rectangle with height 3 and width 11.

This problem is dull because it reduces area and perimeter to simple formulas that never force you to grapple with their original meaning. He notes that here, "'area' doesn't refer to the number of one-by-one squares it takes to cover the rectangle; it's just 'the two numbers, multiplied.'" If you do twenty of these problems, you'll never learn anything about geometry. A more interesting, exploratory problem is at the top of this page.

Hmmm . . . much better. This version requires deeper insight about the nature of rectangles and is way more interesting. Orlin notes that you can take it up another notch with this version: "Create two rectangles so that the first has exactly twice the perimeter of the second, and the second has exactly twice

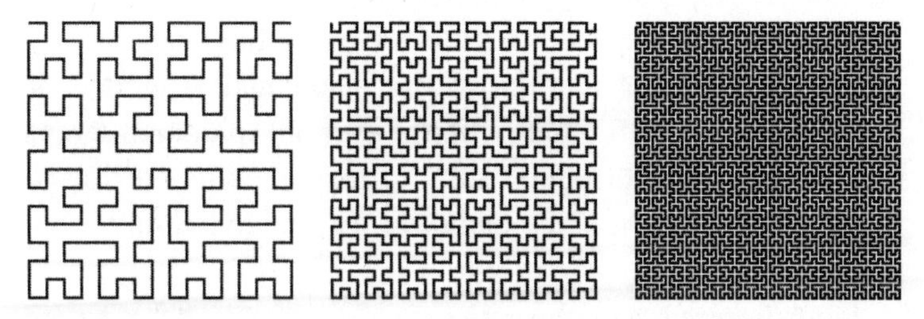

A space-filling curve is constructed as a limit of curves like these, which wind their way through a given region of space in a denser way each time. The mathematical work here is to show that such a limit exists.

the area of the first." Wrestling with good problems like these, you'll invent your own ways of thinking about things and develop your own ways of solving them. This is the best kind of learning.

Exploration cultivates an *expectation of enchantment.* Explorers are excited by the thrill of finding the unexpected, especially things weird and wonderful. It's why hikes through unfamiliar terrain entice us, why unexplored caves beckon to us, why the strange creatures of the deep-sea ocean floor fascinate us—what else may be lurking down there? There's similar enchantment to be found in the zoo of strange discoveries in math. One such curious creature is a *space-filling curve:* a single curve that touches every point inside a square. Although it can't be drawn, only approximated, math tells us this creature exists. As bizarre as they are, space-filling curves now have applications in computer science and image processing.

Another strange animal is the *Banach-Tarski paradox:* the surprising result that a solid ball can be cut into five pieces and reassembled to form *two* solid balls each the same size as the original. You might wonder why this can't be done with balls of gold(!), and the answer—that real matter is not infinitely divisible like idealized space is—helps us grasp the difference between the nature of real things and mathematical models of them. If you go through life with the eyes of exploration, every new landscape is an opportunity to imagine fanciful things, exercise your creative skills, and discover hidden treasure.

Linda Furuto is a mathematics explorer, and she helps others see themselves as explorers too. She grew up on the North Shore of O'ahu in Hawai'i, spearfishing, diving, swimming, and surfing. Although she struggled with math as a child because she didn't see its relevance, she can now see math embedded everywhere around her, from the dynamics of the oceans to the optimization involved in maximizing her time underwater. Now, as a math education professor at the University of Hawai'i at Mānoa, Linda helps students see how mathematics is connected to their cultural histories. She shows students how seeing the world as a math explorer can inform their understanding of marine biology and conservation—the linear functions modeling how coral reefs are being cleared of invasive algae, the matrices describing ocean debris collection, and the quadratic equations involved in sustaining limited island resources. She takes students sailing on *Hōkūle'a* (Star of gladness), a double-hulled canoe of the Polynesian Voyaging Society, on which they learn about the traditional practice of wayfinding among the Indigenous peoples of Hawai'i and the Pacific.[11] Such techniques rely solely on observing clues from nature and the heavens to navigate without instruments. Over the past four decades, *Hōkūle'a* has sailed more than 160,000 nautical miles, including the Mālama Honua Worldwide Voyage (begun in 2013), dispelling any doubts about the reliability of this ancient practice.[12] Linda's role is apprentice navigator and education specialist on land and sea. She helps students explore the trigonometry and calculus embedded in knowing wind dynamics and sail mechanics, and why these are significant beyond memorizing formulas:

> I believe it is important that students know what is written in our textbooks, because they contain important information. However, equally critical is that our students understand and realize that their ancestors sailed thousands of miles across the Pacific Ocean without any kind of modern navigational tool—by the sun, the moon, the stars, the winds, the tides, migratory bird patterns, and more. They traversed oceanic highways in the past, and our students are capable of doing the same things inside and outside of our classrooms today.[13]

Indeed, the wayfinders were mathematical explorers of their society, using attentive study, logical reasoning, and spatial intuition to solve the problems they encountered in their cultural moment. Mathematical explorers have been part of every civilization in every corner of the earth, and Linda sees the importance of drawing the straight line from the mathematical explorers in her students' cultural history to the mathematical identity she'd like them to embrace.

Do you have problems you want to solve? Oceans you want to navigate? Patterns in the starry spheres of your life that you wish to understand? Then you can be a math explorer, since you were born with the human capacities to inquire and to reason. Dream of the sun, the moon, the stars, and the world that you will discover. Imaginative, creative, and unexpected enchantments await.

"DIVIDES" SUDOKU

Sudoku is a puzzle that you solve by exploration. This unusual version is courtesy of Philip Riley and Laura Taalman of Brainfreeze Puzzles, from their book *Naked Sudoku.*[a] There are no numerical clues (it's "naked"), yet it has a unique solution.

Rules: Fill in the cells of the grid so that the numbers 1 through 9 appear exactly once in each row, column, and three-by-three block (the usual sudoku rules). In addition, each time a cell's value exactly divides that of one of its neighbors within a three-by-three block, their common border is marked with the symbol ⊂, whose orientation tells us something: cell A ⊂ cell B indicates that the value of cell A exactly divides the value of cell B. A few "greater than" (>) symbols are also provided.

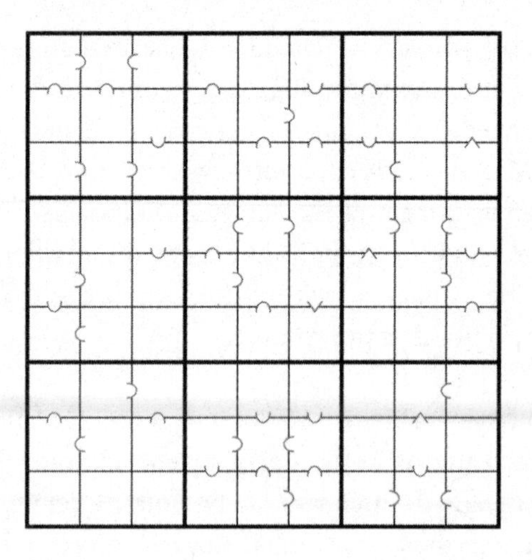

Getting started: You might first think about which numbers divide other ones. For instance, 4 divides 8, so 4 ⊂ 8. Also, 1 divides 3 and 3 divides 9, so 1 ⊂ 3 ⊂ 9. The 1s are usually easy to place, because 1 evenly divides all numbers 1 to 9.

a. Philip Riley and Laura Taalman, Brainfreeze Puzzles, *Naked Sudoku* (New York: Puzzlewright, 2009), 125.

April 16, 2014

Thank you for responding to my letter. I think it was very kind and generous of you to take time out of your busy schedule to read, consider, and respond to my letter, and I appreciated it very much. Yes, I am still continuing to study mathematics: it helps me to pass time productively and gives me a goal to focus on in the immediate to near and distant future. It gives me a personal satisfaction and also a personal hope, because I now know we can only become what we strive for. . . .

My knowledge of mathematics doesn't go far beyond Calculus II. I have never studied in textbooks concerning Number Theory, or Applied Algebraic Systems, or subjects of that nature. I have [an advanced number theory] book here . . . that I strain at, but the gap of me not knowing much about number theory keeps me from getting the enrichment from it that I could. As I grow as a person and a thinker I find myself (seemingly one of the few) to have more of an appreciation of abstraction. And mathematics, being the foundation of science and the basis of our further scientific discovery and technological advancement, is a discipline steeped in abstraction with vast practical implications. The abstraction of mathematics and its importance in all things practical is what has me mystified and curious about mathematics.

Chris

3
meaning

It seems to me that the poet has only to perceive that which others do not perceive, to look deeper than others look. And the mathematician must do the same thing.
Sofia Kovalevskaya

Every word is a dead metaphor.
Jorge Luis Borges

I was skeptical when my dad hired a car to drive our family from the airport to the remote village in China where he was born and where my mother was to be buried. The beat-up clunker didn't look like it would fit the driver *and* the five of

us *plus* all our luggage. I grew more skeptical on the four-hour journey as we wound our way along bumpy dirt paths populated only by goats. Was this a shortcut? Were there really no paved roads to this village?

Then, as we drove over a particularly uneven path, the car's front tires rolled over a bump and the body got wedged against the bump and would not budge. Our front and back tires spun helplessly against the soft dirt. We were stuck.[1]

This wasn't looking good. We were isolated, miles from any real civilization, on a path that was unlikely to receive any traffic. The day was fading fast, and we could not walk dozens of miles before nightfall.

Our predicament didn't seem like a math problem—it had no numbers, no symbols, no formulas—but I couldn't escape the feeling that my mathematical training might help us. I recalled having seen something like this problem before—in a book by the popular mathematics writer Martin Gardner. A truck is stuck beneath an overpass, the problem goes, unable to back up because of traffic, but too tall to proceed forward. What to do? I remembered the answer: let some air out of the tires. This would lower the truck enough to let it roll under the overpass (see next page).

That puzzle seemed somehow similar to our predicament, yet

somehow different—we were stuck not under a pass over us, but over a bump under us. Perhaps we could inflate the tires . . . but sadly, we had no pump. What could we do?

There's a moment in problem solving, as you start to brainstorm and think of potential strategies, when you must make sense of the problem as it really is—when you must strip away its nonessential elements so you can classify it and make connections between this problem and the catalog of problems you've tackled in the past. When you do that, you are wrestling with its underlying meaning.

Indeed, when you want to grasp the meaning of something, you are always asking about its relationship to other things. If you muse about the meaning of life, you are contemplating your place in this world. Or, if you ponder the meaning of a strange event, you have made a choice not to view the event in isolation, but to think about its causes or its implications for other events. And if you look up the meaning of a word, you'll get a definition that places this word in relation to other words.

When the writer Jorge Luis Borges said that "every word is a dead metaphor," quoting the poet Leopoldo Lugones, he meant that every word has a meaning that comes from a certain history—a context in which that word originates. For instance, *calculus* used to mean "small stone," like the kind that you would find in an abacus, to do arithmetic with. Today the word refers to a much more complicated kind of addition. The word *geometry* used to mean "land measurement." Today it refers to the mathematical insights that inform the measuring of almost anything. Words don't exist in isolation. Each one carries with it metaphors from an ancient but ongoing conversation.

Likewise, mathematical ideas, too, are metaphors. Think about the number 7. To say anything interesting about 7, you have to place it in conversation with other things. To say that 7 is a prime is to talk about its relationship with its factors: those numbers that divide evenly into 7. To say that 7 is 111 in binary notation is to have it dialogue with the number 2. To say that 7 is the number of days in a week is to make it converse with the calendar. Thus, the number 7 is both an abstract idea and several concrete metaphors: a prime, a binary number, and days in a week. Similarly, the Pythagorean theorem is a statement relating the three sides of a right triangle, but it is also, metaphorically, every proof you learn that illuminates why it is true and every application you see that shows you why it is useful. So the theorem grows in meaning for you each time you see a new proof or see it used in a new way. Every mathematical idea carries with it metaphors that shape its meaning. No idea can survive in isolation—it will die.

This is why mathematics, like poetry, can be so satisfying.

The meaning of words grows richer the more you use them—they have nuance, they evoke images—so that synonyms are not really synonyms. Poets find great delight in expressing an idea with just the right word. Mathematical ideas, too, grow richer in meaning the more you play with them—each understanding brings a slightly different perspective—so that when you look at an idea in just the right way, you feel enlightened.

Meaning is a fundamental human desire. We crave beautifully written poetry because we enjoy the richness of its meaning. We thirst for meaningful work, if not meaningful lives. We long to connect meaningfully with people. Seeking meaning is a natural expression of living life fully, so why do we settle for less in the way we learn mathematics?

The mathematician Henri Poincaré said:

> Science is built up of facts, as a house is built of stones; but an accumulation of facts is no more a science than a heap of stones is a house.[2]

Learn a bunch of separate mathematical facts, and it is just a heap of stones. To build a house you have to know how the stones fit together. That's why memorizing times tables is boring: because they're a heap of stones. But looking for patterns in those tables and understanding why they happen—that's building a house. And house builders perform better in mathematics; data show that the lowest-achieving students in math are those who use memorization strategies, and the highest-achieving students are those who see math as a set of connected big ideas.[3]

The quest for meaning builds important virtues.

The first is the virtue of *story building*. For millennia, humankind has used stories to convey history or essential truths. A story creates a narrative from disparate events and connects listeners to itself and to one another. It is no different with mathematics. Connecting ideas is essential for building meaning in mathematics, and those who do it become natural story builders and storytellers.

Too often in my math education I've been given a concept and asked to do exercises with it but wasn't taught its significance. I would struggle with the concept, because even though it had a definition, it had no meaning—no connection to a larger narrative. However, there were many instances when a story, as captured by a pithy phrase, helped me see the bigger picture. In calculus, when someone said, "Integration by parts is the inverse of the product rule," both concepts became clearer. In statistics, I heard this story: "Learning statistics is learning to be a good detective with data." And in all of mathematics there is this lesson: "Objects are not as important as the functions between objects." This maxim recapitulates what I am saying about mathematical meaning: objects don't have meaning independent of their relationships to other objects. Functions are relationships; functions tell a story.

Now, there are many ways to build a story. Think again of the Pythagorean theorem, which says that the side lengths a, b, and c of a right triangle (a triangle with one ninety-degree angle) satisfy the relationship

$$a^2 + b^2 = c^2$$

where c is the side length of the hypotenuse (the longest side). This is a fact without context and easily forgotten, until you build a story.

You might create a *geometric story*, by drawing the right triangle with squares on each side and realizing that the theorem says the areas of the two smaller squares must add up to that of the largest square (see next page).

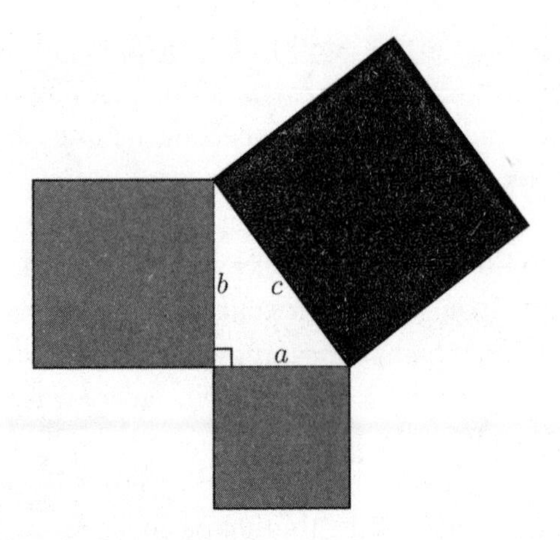

You could ask for a *significance story,* to explain why it matters: "The Pythagorean theorem is the basis of all trigonometry and one of the most important theorems in geometry." A *historical story* would situate the theorem in a context in history: "The Pythagoreans' proof of their theorem was discovered a couple of centuries before Euclid's proof of the theorem."

Math explorers like *explanatory stories.* That's what a proof is. The figure on the facing page demonstrates what's called a "proof without words"—a diagram showing the truth of the Pythagorean theorem by dissecting squares into pieces. The corresponding pieces show how the area of the larger square must be the same as the sum of the areas of the smaller two squares. (You'll still want to ponder why this kind of dissection will work for *any* right triangle.)

You might be surprised to learn that there's a *physical story* for the Pythagorean theorem. If you write the velocity vector of an object as a sum of motions along its horizontal and vertical components, you'll get a right triangle of vectors. Because speed is the length of a velocity vector, and kinetic energy is propor-

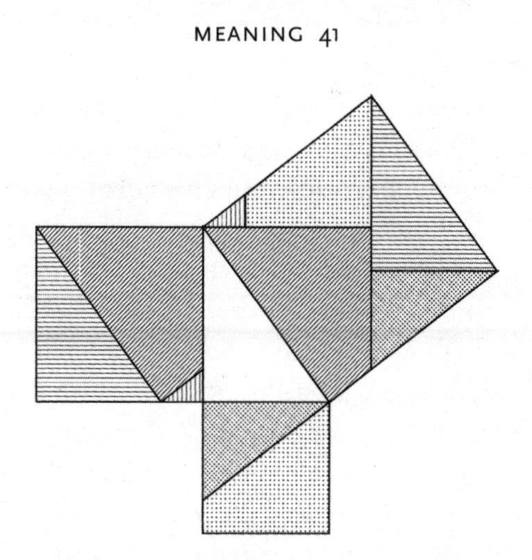

tional to the square of the speed, the Pythagorean theorem says that the energy needed to push the object in a diagonal direction is the sum of the energies needed to push it first in the horizontal direction and then in the vertical direction.

You can create an *experiential story* through play, inquiry-based learning, or building physical models. Try this carpenter's trick for making a right angle with two beams of wood: since $3^2 + 4^2 = 5^2$, you can place two beams together at a corner, make a mark 3 units from the corner on one beam, make a mark 4 units from the corner on the other, and then adjust the angle so that the two marks are exactly 5 units apart. Then you'll know the corner angle is a right angle.

Each of these stories adds to the meaning of the Pythagorean theorem for you. Stories are an essential part of retaining new knowledge. It is much easier to remember things when they make sense in a story.

A great example of using multiple kinds of story building to reinforce learning can be found in the Algebra Project, a math literacy effort in the US for economically disenfranchised communities. The project, founded by the MacArthur fellow and civil rights activist Robert P. Moses, provides curricula and training for teachers and reaches nearly 10,000 students per year. It uses experiential learning, with five steps moving from experience to abstraction: (1) Physical Events, such as taking a trip or making an observation; (2) Pictorial Representation/Modeling, which asks students to draw a picture or build a model of the physical event; (3) Intuitive Language/People Talk, which asks students to tell stories about the physical event; (4) Structured Language/Feature Talk, in which students isolate features of the event that can be studied with math; and finally (5) Symbolic Representation, where students build models of their ideas. Notice how each step is a form of story building.[4]

A second virtue that a quest for meaning builds is *thinking abstractly*. People often think of abstraction as stripping away meaning. But in fact, abstraction does the opposite—it enriches meaning. When you see that two things have similar structures or behavior, then those similarities create a connection, a new meaning for you that wasn't there before. Poincaré famously said that "mathematics is the art of giving the same name to different things." (To which a poet quipped: "Poetry is the art of giving different names to the same thing.")[5] If you've only ever seen one dog, you might think that a dog must be a German shepherd. Once you've seen several, you begin to realize that the meaning of *dog* is richer than you realized. Abstraction enriches meaning by helping you to take a collection of examples and to see just what is essential about, for instance, "dogness." In so doing, you see what's the same about many different things.

Thinking abstractly is one of the main benefits of learning algebra. It's easy to get so caught up in practicing fluency with algebra skills—manipulating expressions, factoring—that you may fail to pause and appreciate algebra's larger power: building flexible thinkers who can recognize patterns in relationships and reason in general ways to solve a host of problems at once. Using algebra, we build general formulas for compound interest, or calories burned, or coin-toss probabilities, so that they will work in many different situations, not just the one we are facing right now. Viewed one way, the quadratic formula is just a formula useful for solving quadratic equations. But viewed another way, the quadratic formula makes many different problems look the same. Abstraction promotes flexible thinking, a necessary skill in almost any profession. The skill of deriving a formula that works in many situations will translate into the skill of writing a flexible computer program that can handle any input, or designing a building that works for many different people.

The skill of abstract thinking pays dividends, not just in careers but in other areas of our lives. Don't we need to strip away extraneous details of a problem and look for its essential nature? Don't we want to look at problems from multiple points of view? Aren't we better equipped to do this because of mathematics? So when I hear the puzzle about a truck going under an overpass, I don't dwell on extraneous details, like the fact that it's a truck or how much it weighs. I strip away all the details, and I attempt to see what's essential to that puzzle by looking at it from many different perspectives. In so doing, I prepare myself to recognize when a future problem—like a car stuck on a bump—is essentially the same problem as this one.

A quest for meaning produces the additional virtues of *persistence* and *contemplation.* Discerning the meaning of an idea takes sustained thought. This is the hard work of problem solving, when you must sit with a math problem and meditate on it. In doing so, you form mental connections and build stories to explain the patterns you see. William Byers, in his book *How Mathematicians Think*, gives numerous examples of ideas we take for granted as "easy concepts" once we've learned them that are actually quite deep if you reflect on their meaning.[6] For instance, in the equation $x + 3 = 5$, the left side represents a process (addition), but the right side is a number. How is it that we are equating a process with a number? Similarly, when you first try to solve this equation, the variable x could stand for *any* number, but by the end it is only one number: $x = 2$. So which is it: any number or one number? Resolving these ambiguities is the key to understanding the meaning of the expression, and that requires contemplation. In the end, the resolution results in joy. With accumulated joy from prior successful quests for meaning, we cultivate persistence—in hopeful anticipation of further reward.

So, at the heart of a proper practice of mathematics is a quest for meaning. You cannot flourish in mathematics or in life without finding meaning in your endeavors. I like this definition of *mathematics:*

Mathematics is the science of patterns.[7]

But I would add a reflective component to this definition, because the term *science* here makes it sound like we do math only to produce discoveries. In reality, there's more to math than its utility, and beauty is found by reflecting on an idea's many meanings as we swivel from perspective to perspective. So I prefer to say:

Mathematics is the science of patterns
and the art of engaging the meaning of those patterns.

Our car was still stuck on that bump.

While others resigned themselves to a long night in the car, I continued to contemplate our problem, because my mathematical persistence, fortified by past struggle with meaning, kept me from giving up.

My mathematical story building had placed the puzzle of our predicament in the category of the puzzle about the truck stuck under the overpass.

My mathematical penchant for abstraction had stripped our puzzle down to its barest essence: a problem that at first appeared to be about the relationship between a vehicle and a bump but was really about the relationship between a vehicle and its tires. How, then, should I think about our vehicle and its tires if we wanted to raise our car but had no pump to inflate the tires?

That's when the insight came. *Get out of the car.*

After being unburdened of five people and baggage—about 700 pounds—the cab of the car lifted off the bump, and we could roll the car forward. We were free.

RED-BLACK CARD TRICK

Here's a pretty simple, yet surprising, card trick, as it appears to others: You give a deck of cards to a spectator and ask her to shuffle it and return it to you face down. You take the cards and (with a little showmanship but without looking at their faces) separate them into two piles, and then say, "The number of red cards in the first pile is the number of black cards in the second pile." Have your spectator turn over the cards and verify!

Here's how to do the trick: A standard deck will work, but to reduce performance time it's best to use a deck of about twenty cards with equal numbers of black and red cards. When the spectator gives you the shuffled deck, all you need to do (though don't make it obvious) is to count the cards into two piles so that there are an equal number in each pile. Can you figure out why this works?

This trick and the following related puzzle appear in Ravi Vakil's delightful book *A Mathematical Mosaic.*[a] Can you see how the two are related?

WATER AND WINE

Take two glasses of the same size and pour wine into one until it is half full, and pour water into the other until it is half full. Now take a spoon of wine from the first glass and put it into the second glass, then mix it around. Without worrying how well the wine has mixed in, take a spoon of the liquid from the second glass and put it into the first.

Is there more wine in the water glass, or more water in the wine glass?

a. Ravi Vakil, *A Mathematical Mosaic: Patterns & Problem Solving* (Burlington, Ontario: Brendan Kelly, 1996).

August 9, 2018

When you speak of "the art of making meaning with patterns," the "artistic nature of math," the "meaning in patterns" creating meaning, making symbols stand for something, and "picking the way of seeing," those seem like poetical ways to see and describe mathematics. That's how I want to see and describe mathematics. That resonates with my experience, because I can see it somewhat. When the British mathematician said about Fourier, "Fourier is a mathematical poem," that's part of my goal, to see, understand, and describe Mathematical Poetry. Not only chiefly in math, but also over to how I view all areas of life. And I think I understand what you're saying. It's like the creating of chess, where we take symbols (pieces) and give or create in them meanings (rules) and then see the relationships and interactions among them which create that universe or circumstance among them. Or the creation of Non-Euclidean geometry also, with its symbols and meanings that create its own universe or circumstance.

Chris

4
play

Play is the exultation of the possible.
Martin Buber

It matters little who first arrives at an idea, rather what is significant is how far that idea can go.
Sophie Germain

Just beyond the first breaths of life, beyond basic needs that must be met, a baby begins to understand her world through play. She coos and waits for a parent to coo back. She kicks her legs in random patterns. She puts fingers in her mouth and explores the strange resulting sensation from two points of view:

her fingers and her mouth. She is driven not by need but by curiosity. She is playfully probing her surroundings through call-and-response, pattern explorations, and a change of perspective. She is beginning to embody mathematical modes of play.

As she grows, her desire for play changes character. It becomes communal. It begins to shape the way she learns, works, and interacts with others. It takes on cultural expressions, manifesting itself in various ways—sports play, musical play, wordplay, puzzle play. Facets of play run through nearly every human activity: dancing, dating, crafting, cooking, gardening, and even "serious" activities like work and trade. Indeed, the cultural historian Johan Huizinga argued that play has had a strong influence in shaping the archetypal activities of civilized society: language, law, commerce, art, and even war—each of these has elements that derive from play.[1] Along similar lines, the writer G. K. Chesterton said, "It might reasonably be maintained that the true object of all human life is play."[2]

Play is a deep human desire, and it is a mark of human flourishing.

What, then, is play? Play is hard to define, but it has several characteristic qualities. If play is an activity engaged in for enjoyment or recreation, then play should be *fun.* But this definition doesn't tell us why play is fun. Babies love to play peekaboo, but why?

Other characteristics of play shed more light on its nature. For instance, play is usually *voluntary.* If you force me to perform piano scales over and over, that might be good practice, but it would not feel playful. Play is *meaningful,* or we wouldn't engage in it. Play follows some *structure.* Think of the rules of a game, the arrangement of musical chords, or the pattern of peekaboo. Play inhabits the *freedom* within that structure—like the choices of strategies in a game, the rhythm of music, the moves of a Rubik's cube, or where I appear in peekaboo. That freedom leads to *exploration* of some sort—the hunt to solve the Rubik's cube, the game play in football, or improvisation in jazz. The exploration can often lead to some sort of *surprise*—a solved cube, a "peekaboo," a joyous musical riff, a satisfying pun in wordplay, or a thrilling end to a football game. Of course, animals play too, but what characterizes human play is the enlarged role of the creative mind and the *imagination.*

As Huizinga noted, play often pulls its participants out of their usual life and into "a temporary sphere of activity with a disposition all of its own."[3] Children play "pretend"; adults gather around the card table for a game; a dance number captivates us for three minutes. To borrow a term from music, we may describe play as an *interlude* in the symphony of life, or, to borrow a concept from computing, it is a *subroutine* in the routine of life. This interlude or subroutine establishes its own world of existence, is engrossing for the participants, and results in a temporary *absorption,* or focused attention. For that reason, play sometimes feels like escape. In the best forms of play, the players *do not ridicule one another* for their performance, they *respect and give one another ownership* of the play, and there is *no long-term stake in the outcome*—you might care if you win a game while playing it, but next week it won't matter.

Mathematics makes the mind its playground. Doing math properly is engaging in a kind of play: having fun with ideas that emerge when you explore patterns, and cultivating wonder about how things work. Math is not about memorizing procedures or formulas, or at least that's not where you start. It's the same way in sports. In football, you wouldn't practice drills unless you wanted to play competitively, but you can start with an enjoyment of the game.

Sometime in my youth, I was introduced to a nifty short-

THE GAME OF CYCLES

Since we're thinking about play, let's invent a new game and explore it. Grab a friend and play this game. Draw this starting diagram (below) of dots and edges, which divides a triangular region into three smaller triangles, called *cells.*

Players take turns, marking a single arrow along an edge of the diagram, obeying these rules: only one arrow may be placed on each edge, and no dot can be a sink or a source. A *sink* is a dot that has all its adjoining edges marked with an arrow pointing toward it. A *source* is a dot that has all its adjoining edges marked with an arrow pointing away from it. The sink-source rule means that some edges may become unmarkable during the course of play.

The object of the game is to produce a *cycle cell*—a cell bordered by arrows that cycle in one direction, clockwise or

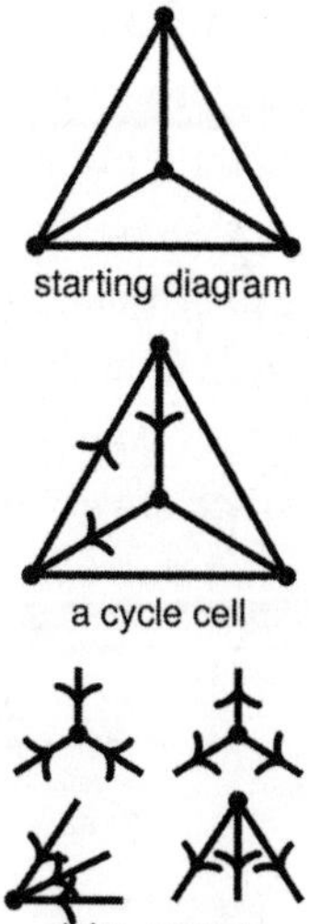

starting diagram

a cycle cell

sinks, sources not allowed

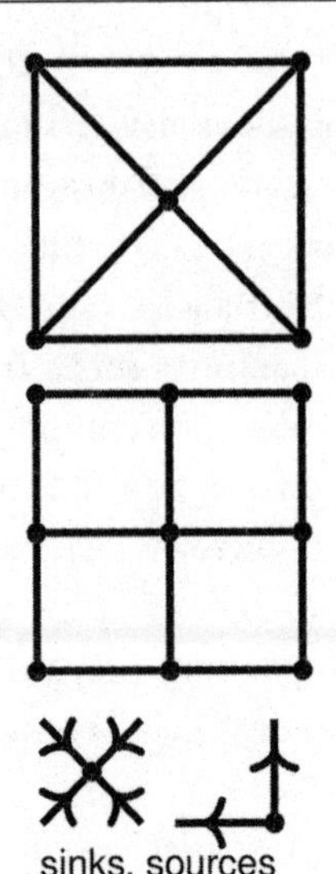

counterclockwise. The player who creates a cycle cell or makes the last possible move is the winner.

After you play this game for a while, see if you can figure out whether the first or the second player has a winning strategy. Then explore the game of cycles on these other diagrams above. Note that cells don't have to be triangles.

I just created this game, and at the time of writing I'm unaware of whether it has been invented or studied before. So there are lots of open questions, and I'm playfully exploring the game, just like you!

cut for squaring a number that ends in 5. This shortcut is easy enough to do in your head. If you are curious, you might try to figure out what it is. It may surprise you to know that there is a pattern at all, so you may not have thought to look, but a seasoned math explorer begins to expect enchantment—to believe that patterns abound everywhere, waiting to be uncovered. Indulge your curiosity and try a few examples!

That's what doing math looks like when you learn any new idea—you play with it. Even for professional mathematicians, the beginning of a research project is playful exploration: contemplating patterns, playing with ideas, exploring what's true, and enjoying the surprises that arise along the way. In collaborative research, there is no judgment when one makes a false start; in fact, that's part of the fun of exploration. Math explorers pursue not only problems that have immediate application but also questions without long-term stakes that are intrinsically appealing. There's even a whole area of mathematics known as recreational mathematics. Is there another academic discipline with a recreational subfield? (I suppose there's "recreational chemistry"—but you should probably stay away from that.)

Mathematical play showcases some refinements of the characteristics of play. Math play is voluntary, but driven by a deep curiosity that is nurtured through practice—much like how the more you get a feeling for a new game, the more you want to play. The most practiced adherents of math play will see a new idea and feel a compulsion to play with it. The structure and the freedom of math play must, of course, follow the laws of mathematics. Beyond this, the rituals of math play often happen in two phases of mathematical exploration.

The first phase is the inquiry phase, when an explorer engages in *pattern exploration* and uses *inductive reasoning* to make general claims from specific instances—these are called conjectures. The results of play frequently produce interesting patterns. For instance, if you explore squares of numbers ending in 5, you'll find:

$15^2 = 225,$
$25^2 = 625,$
$35^2 = 1225,$
$45^2 = 2025.$

In math education, it has become fashionable to show students some mathematical object and ask them: "What do you notice? What do you wonder?" For the squares we just computed, do you observe a pattern?

Such questions invite us to deeper reflection on what we see. They initiate a rich dialogue that can take place between the explorer and the mathematics. I like what Paul Lockhart says in "A Mathematician's Lament": "This is the amazing thing about making imaginary patterns: they talk back!"[4]

This dialogue is a version of the call-and-response found in so many other kinds of play, in which an action generates a reaction. You hear it in musical forms like jazz when one instrument responds to another, or in military marches when the leader calls out a line of a song and his unit repeats it, or in a back-and-forth rally of a tennis match, or in babies cooing and parents responding.

In mathematical play, call-and-response happens when the mathematics calls out to the explorer and asks, "What do you notice? What do you wonder?" and the explorer responds with an observation: "The squares always end in 25. I wonder what the numbers 2, 6, 12, and 20 have in common." This dialogue happens several more times, until the explorer has enough observations to make a claim. For instance, after looking at the list of squares, you may respond in delight, "I see a pattern!" If so, you'll have a conjecture.

Then the call-and-response reverses direction, as the explorer tests her claim by trying an example and the mathematics responds by producing a confirmation or refutation of that claim. For instance, in the squares problem, once you have a conjecture to test, you'll call out to the math, "Show me 55^2, please" (yes, politeness is appropriate here), and the math will respond, "3025." You'll check if your idea works, and then you'll try again: "Now show me 65^2, please," and the math will respond, "4225."

This cycle may happen several more times before an explorer decides that her conjecture is solid. Along the way she may realize the importance of a certain concept, and clarify that idea with a definition. You see, in the interlude of math play, she has the power and the creative freedom to establish new norms.

Thus math play is like a baby cooing and listening for a response. Math play is like a preacher speaking truth and waiting for an "Amen." Math play is like a tennis player in a game against a new opponent, trying out different shots to see how he responds.

The second phase of mathematical exploration is the justification phase, when one engages in *deductive reasoning* to provide a logical explanation for a conjecture, by providing a proof or a mathematical model: a description of what's going on in mathematical language. There are some well-worn modes of math play in this phase. To show that a statement is true, a math explorer might attempt what's called a *proof by contradiction*—assuming it's not true and reasoning toward a contradiction. This rules out the possibility that the statement is false. To show the truth of all the statements in a sequence, an experienced math explorer might try a *proof by induction,* which uses the truth of one statement to deliver the truth of the next, the way a line of dominoes fall when you tip the first one. Such a collection of opening moves for a math proof is just like a catalog of opening moves in chess. They'll get you off to a good start.

For mathematical models, a useful pattern of mathematical play is to make simplifying assumptions. You are changing the scope of your play to make the problem easier to solve. For example, to construct a mathematical model of cooling coffee, an experienced math explorer might make some simplifying assumptions about the liquid and its rate of cooling, understanding that the simplification may not capture all the features of the problem but will retain the most salient ones.

Another pattern of mathematical play is the way it asks you to change perspective, to look at a problem from different viewpoints. Once, when I entered a dark cave with a tour group, our guide asked us to turn off our lanterns and experience the cave as a bat would, with no light, just sound. I yelled and listened for echoes. This change of senses gave me a new way to perceive the cave. In mathematical play, changing perspective is an essential feature of problem solving. You will notice different aspects of a problem when you probe it in different ways, looking from different vantage points, and you will have multiple strategies for finding a solution. It is for this reason that mathematicians sometimes doodle—drawing diagrams to represent complex relationships—even if the problem they are thinking about has no spatial component. Or they will choose a different notation or definition. As one of my students once observed, "Choosing a good notation or definition is like specifying what kind of conversation you are going to have with the material." This skill of changing perspective equips teachers with multiple ways of explaining a concept to others.

Thus, math play is like a chess player who scans her experience to pull out an opening gambit. Math play is like a hunter who picks an appropriate arrow from her quiver. Math play is like a cook who chooses a good spice for his culinary creation. Some choices are better than others, but many choices will work, and they will yield different dishes.

The conjecture responds by yielding to these maneuvers or responding with obstacles and new challenges. The creativity of a math explorer is tested here, because she must constantly pull weapons from her arsenal to respond to whatever obstacles the conjecture may put up.

We've spoken of two phases of mathematical exploration, but in reality these are artificial distinctions, because when the second phase is complete and the proof or model has been established, new questions are generated that suggest new things to prove or a refinement of the model to explore. So it may be more appropriate to think of exploration as a continuous cycle passing from one phase to the other and back again.[5]

We can illustrate the cycle of math play by returning to our problem of finding a shortcut for squaring numbers that end in 5. Once you have a well-tested conjecture, you can move to phase two: trying to prove *why* this shortcut works. If you know some algebra, you might set up a general expression for numbers that end in 5 by noting that any such number has the form $10n + 5$ for some number n. Now see what happens when you square it. Does your expression confirm your conjecture? If so, you've just showed that this shortcut works for *every* number that ends in 5. I'll leave this to you as an opportunity to have your own moment of discovery.

Unlike you, I wasn't given the opportunity to explore and discover this trick on my own—someone just showed me, so I wasn't able to have an *aha!* moment. Quite often we treat mathematics as a thing you just *tell* people, even though, as the French philosopher-mathematician Blaise Pascal reminds us, "People are generally better persuaded by the reasons which they have themselves discovered than by those which have come into the mind of others."[6]

For me, learning this cool squaring trick was a springboard for further investigative play. I began to wonder: If numbers that end in 5 have squares that end in 25, what can be said about numbers ending in 25? A small challenge to you, if you choose to accept it, is to figure out the rule, then prove it.

A second observation is far more interesting. When you multiply two *different* numbers ending in 25, the result always ends in 25. That's kind of a neat property, so let's give it a name. It is a fun and creative aspect of math play that we get to name the ideas we invent inside our universe.

Let's call the final digits of a number an *ending*. Then call an ending *stubborn* if the product of two numbers with that ending must also have that ending. So . . . 25 is a stubborn ending. This invites the question

> Which other two-digit endings are stubborn?

Note how easy it was to state our question, because we made good definitions. After some exploration,[7] you will find just four stubborn two-digit endings:

. . . 00,
. . . 01,
. . . 25,
. . . 76.

You probably could have guessed that the . . . 00 ending would be stubborn (why?), but the others are not as obvious. It certainly surprised me to learn that the product of any two numbers that end in 76 will also end in 76. How about three-digit endings—which ones are stubborn? Will I need to test all 1,000 possible three-digit endings to find out? (Hint: no.) Again, you find only four:

. . . 000,

. . . 001,

. . . 625,

. . . 376.

Which four-digit endings are stubborn? Which five-digit endings are stubborn?

As I worked my way through this series of questions, each answer surprised me. For any given length, there only ever seem to be four stubborn endings that "continue" a pattern from the previous length ending. So

. . . 625

led to

. . . 0625,

a four-digit stubborn ending, and that led to

. . . 90625,

a five-digit stubborn ending. If you keep on going, you'll find that

. . . 259918212890625

is the unique fifteen-digit stubborn ending that ends in 5. It is the tantalizing ending of a mysterious string of numbers that has no beginning! I felt that a hidden beauty of the mathematical world was revealing itself.

What happens to the other stubborn endings when you

lengthen them? If you wish to explore, and not be told what happens, do not read the note to this sentence.[8] Just try it. What I found was so fascinating that I computed all fifteen-digit stubborn endings, and then I asked another question: What is the relationship between these number endings?

I sat and stared for a while.

And then, in a blinding moment, I saw a pattern! (You do not need to go out to fifteen digits to see this pattern, by the way.) I was in numinous awe, as if the universe had opened up and shown me something profound. I wondered if anyone else had ever seen this. I wanted to share my excitement with another human being. It was a thrilling, beautiful pattern, yet mysterious as well! I did not know why it was true, though I felt it *must* be true.

That pattern would not reveal its secrets until a few years later, when I took a number theory course in college and learned about the Chinese remainder theorem. I was finally able to see why there was a pattern, and prove it. I also learned later that stubborn endings had already been investigated by others,[9] but that didn't matter. The joy of play was in seeing how far these ideas could go.

You don't need to know number theory to marvel at what just happened. Just by noticing the pattern and asking the question *why?*, you are engaging in mathematical play. You have left your worldly rhythm and absorbed yourself in a separate interlude, one that has rewarded your persistence and play with surprise, joy, and a deeper connection to what is true. You have gained some skills in a new dimension that will help you grow in a new way.

You see, the proper exercise of mathematical play builds virtues that enable us to flourish in every area of our lives.

For instance, math play builds *hopefulness*—when you tinker with a problem long enough, you are exercising hope that you will eventually solve it. This experience of hopefulness carries over to other hard problems we wrestle with. Math play builds *curiosity* as you explore, and it develops *concentration,* an intensely pleasurable focus that shuns the distractions of daily life. Simone Weil said:

> If we have no aptitude or natural taste for geometry, this does not mean that our faculty for attention will not be developed by wrestling with a problem or studying a theorem. On the contrary it is almost an advantage.[10]

Math play builds *confidence in struggle*—you know what it's like to struggle, because you're used to it, and you welcome it, and you realize that if your brain hurts from the struggle that it is a welcome working of the mind. Math play cultivates *patience* in the discipline of waiting for a solution, possibly several years, until you see its conclusion. Math play builds *perseverance*—just as weekly soccer practices build up the muscles that make us stronger for the next game, weekly math investigations make us more fit for the next problem, whatever it might be—even if we don't solve the current problem.

And math play builds the *ability to change perspectives,* to see a problem from many viewpoints, as well as an *openness of spirit* that contributes to community—when you share in the struggle and the delight of working on a problem with other human beings, you begin to see them differently. These are some of the most important virtues that we cultivate through math play.

Openness of spirit is also one of the virtues that are most easily tarnished when play is corrupted. For instance, an undue emphasis on performance can lead to an unhealthy competitiveness. Elevating the stakes can spoil the fun and the openness of play. The nobility of play can also be marred by an exclusiveness, born of arrogance, about who is able to play.

The mathematician G. H. Hardy wrote a well-known defense of mathematics, called *A Mathematician's Apology,* that is spirited, and at times effective.[11] However, in it he seems to prize mathematical achievement as the most important goal of a mathematician—to such an extent that he speaks mockingly of "trivial" problems and passes judgment on the triviality of his own mathematical contributions. To me, this is a misguided overemphasis on the stakes of doing math, which sadly strips the joy from what should be an eminently playful enterprise.

We would teach math so differently if we thought of it as a playful sport, not a performance sport.

There's a nuanced conversation to be had about math competitions, a form of math play that encourages community through a shared experience in problem solving. I know some question their value, and I can understand why. Such gatherings sometimes encourage an unhealthy competitiveness, especially if the competition is not well designed, or if it rewards speed (a computational skill, not a mathematical skill) over ingenuity. Also, competitions often attract a limited subgroup of those who might enjoy mathematics more broadly—and some are never invited. This has the unfortunate effect that the public thinks winners must be representative of anyone who is "good at math." That's a strange association. Think of all the gifted athletes who were never interested in or never tried the hundred-meter dash. It would be silly to call the hundred-meter dash a "sports competition" and consecrate its winners as the only athletes who are "good at sports."

Nonetheless, I've seen many benefits of well-designed math competitions. I've seen kids who, for the first time, have found community and camaraderie among others who share the same passion for interesting problems. They aren't teased or shamed in these communities for any quirks they might have. They are accepted with an openness of spirit by all who are there. And if the problems they tackle are interesting enough, the math play will continue afterward as these newly formed friends discuss their ideas.

In 2016, a team from the United States won the International Math Olympiad for the second year in a row. This was a marked achievement for the US team, since prior to 2015 it had not won for two decades. What received less attention was the fact that the mathematician Po-Shen Loh, who coached the US team, had invited teams from other countries to train with them to prepare. He prioritized community over competition. He emphasized the joy of solving problems together. This action was so impressive to the Singaporean prime minister that he publicly thanked President Obama for this remarkable collaboration.[12] Winning was not as important as enjoying the true spirit of mathematical play.

We can draw ourselves and others to mathematics through the deeply human desire to play. Thus, play should play a prominent role in mathematical learning. Everyone can play. Everyone enjoys play. Everyone can have a meaningful experience in mathematical play. As Plato said: "Do not, then, my friend, keep children to their studies by compulsion, but by play."[13]

There are many ways to make math play central to your own learning experience. Look for and expect patterns all around you. Whenever you encounter one, start asking questions. If you're given a problem, play with it to get a feeling for it before you look for an answer. Whenever you successfully solve a problem, practice asking a follow-up question that leads to further investigation. Build a community around you—in your home, classroom, or friendship circle—that values asking interesting questions. As a parent or as a teacher, you might find this scary, because the activities may generate questions that you can't answer. But that is part of modeling who an explorer is: you won't always know the answer, but you will know how to highlight and cultivate the virtues, built through math play, that will help others find the answers they are looking for.

Play is fundamental to who we are as human beings, and the desire for play can entice everyone to do, and enjoy, mathematics.

A GEOMETRIC PUZZLE

These three overlapping rectangles are congruent (they have the same size and shape), each with area 4. The black dots mark the midpoints of the short sides. The three rectangle borders meet at one point in the middle. What total area is covered by this design?

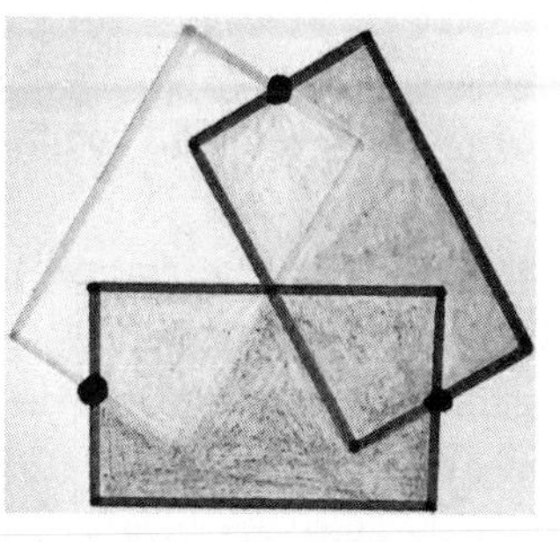

This nice puzzle was created by and provided courtesy of Catriona Shearer, a mathematics teacher in Cambridge, UK. She enjoys making geometric puzzles like these, which have become hugely popular on Twitter. Her hobby began as an exploration! In her words:

> I went on holiday to the Scottish Highlands, but forgot to take a coat with me, so I ended up spending more time inside than my friends did! I kept doodling along the lines of "I wonder if I could work out . . ."
>
> I wasn't expecting it to turn into a hobby, but it gets a bit addictive. . . .
>
> It just starts with doodling. I'll end up with a whole page of overlapping squares at different angles, or regular(ish) pentagons with different parts shaded in, and then see if there's any nice Maths hiding there—relationships between lengths or areas or angles.[a]

a. See Shearer's interview with Ben Orlin in "Twenty Questions (of Maddening, Delicious Geometry)," *Math with Bad Drawings* (blog), October 3, 2018, https://mathwithbaddrawings.com/2018/10/03/twenty-questions-of-maddening-delicious-geometry/.

Spoiler alert: Chris solves the "Dividing Brownies" puzzle (from chapter 1) in this letter, so skip to the final paragraph if you don't want to read the solution.

January 28, 2018
Francis:

I'm fairly confident I solved your problem correctly this time. At first this problem seems highly conditional. So here are 2 more of my conditional answers. . . .

2(a). If the rectangles don't share a center but one of the diagonals of the outer rectangle intersects the inner rectangle's center, a cut along this said outer rectangle diagonal will do.

2(b). If the center of the inner rectangle is equidistant from both of the lengths of the outer rectangle, a horizontal line going through the inner rectangle's center will do.

But then I got to thinking about slopes of lines, the triangles inside the inner triangle, the areas of the inner triangles, hypotenuses, legs, side-angle-side, distance, and then a line is determined by the distance between 2 points, and a straight line can be drawn through any 2 points.

Then it hit me in the head.

3. (General answer): In short, the diagonal line that goes through both the center of the inner rectangle and the center of the outer rectangle will suffice.

That was a good problem: it was instructive, and it reminds you to think of the things that you already know. After I sent you my first answer . . . all of a sudden it hit me that part of my answer wasn't exactly accurate. I had the feeling of how to make it more accurate, but that would've clearly overcomplicated it. Thank you for pushing me to a better answer.

Chris

5
beauty

The yearning for and the satisfaction gained from mathematical insight bring the subject near to art.
Olga Taussky-Todd

Why do you want to share something beautiful with somebody else? It's because of the pleasure [s]he will get, and in transmitting it you appreciate its beauty all over again.
David Blackwell

When I was in college, one of my general-education requirements was to take an art class. I'll admit I didn't have an excitement for art at the time. A friend suggested that we take an "ar-

chitecture appreciation" course because it was easy and because we could spend most of our time slouched in a cool, dark auditorium looking at pictures. I had no expectation that I would be inspired. Yet my experience was, in many ways, life changing. The professor took us on a tour of beautiful buildings, and helped us to see why they were so revered. Some were obviously masterpieces, and some took a while to appreciate. I began to see a relationship between form and function. I began to see historical trends. I began to recognize what I personally liked and what I didn't care much for, and to understand why. I began to value the culture and context around architectural beauty.

Since that class, I have never looked at a building the same way again. I can now often look at a building and tell you the era in which it was built. I can often imagine what the architect was trying to accomplish. I remember that feeling when I first walked the Harvard campus and saw Sever Hall and was able to recognize it as the work of a particular architect, H. H. Richardson, even though I'd never seen the building before. It was an *aha!* moment, made possible because I had been sensitized to architectural beauty.

Maryam Mirzakhani, who won the Fields Medal (the world's top prize for younger mathematicians), once said: "The beauty of mathematics only shows itself to more patient followers."[1] She was noting that mathematical beauty can sometimes take a while to unfold. Just as architecture sometimes takes some effort to appreciate, mathematical beauty sometimes rewards patient followers with a slowly revealing grandeur.

But the beauty of mathematics can also reveal itself to a math explorer in a blinding instant, when she suddenly sees an elegant solution to the problem she's been wrestling with. That's the *aha!* moment, in which all the puzzle pieces come together and everything becomes clear. Just like my experience with Sever

Hall, a mathematical epiphany is the thrill of recognizing something profound.

I believe that many of you have experienced mathematical beauty but never realized it. It's like you've appreciated the buildings around you without looking at them more deeply. You've seen the function of the buildings without appreciating their form. I also realize that a few of you may never have seen buildings before, and that's okay too—for you, it may take time to get used to buildings before you decide which ones you admire. So, if you've truly never beheld mathematical beauty, I will help you grasp what all the fuss is about, to help you embrace the world as the math explorer you are.

The desire for beauty is universal. Who among us does not enjoy beautiful things? A striking sunset. A sublime sonata. A profound poem. An illuminating idea. We are drawn to beauty. We are enamored of it. We dwell on it. We seek to create it. Beauty is a basic human desire, and expressions of beauty are marks of human flourishing.

Beauty appears in mathematics, too—in many forms—though it is less appreciated there because many have not had a chance to experience it, or they have experienced it but not recognized the experience as mathematical. But math explorers and professional mathematicians usually cite beauty as one of the main reasons they pursue mathematics. One study even showed that mathematicians respond to mathematical beauty in a similar way to how others respond to visual, musical, or moral beauty—with the activation of a part of the brain that handles emotion, learning, pleasure, and reward.[2]

Thus, if you've ever enjoyed a sunset, sonata, or poem, wouldn't you also want to try experiencing mathematical beauty? It's within your reach.

Many have tried to define or characterize the general nature of beauty. And philosophers of aesthetics have debated to what extent it is subjective (dependent on the observer) or objective (depending on qualities inherent in the object being admired). I will not attempt to resolve that debate here, because it's sufficient for my purposes to admit that it's not entirely one or the other. On the one hand, we can't deny that individuals have different tastes and that culture influences what's regarded as beautiful. On the other hand, there are characteristics of mathematical beauty on which mathematicians are in wide agreement. Many have tried to write down such lists: among them, the mathematician G. H. Hardy argued that the beauty of a mathematical idea often rests on its "seriousness" as well as its unexpectedness, inevitability, and economy of expression; the philosopher Harold Osborne surveyed writing on beauty in mathematics and summarized the qualities named there as order, coherence, lucidity, elegance, clarity, significance, depth, simplicity, comprehensiveness, and insight; and in his book *How Mathematicians Think,* the mathematician William Byers makes a strong case for ambiguity, contradiction, and paradox as important features of mathematics that some would consider beautiful.[3] But listing characteristics of beauty for the uninitiated is a lot like explaining the allure of good sushi by citing its temperature, texture, or acidity.

It may seem a hopeless task to try to describe to you something you may not have experienced, like describing a rainbow to someone who has never seen color. But in fact, the blind have sought to "sense" color by attaching it to other senses or emotions. So I reject the pessimism of the mathematician Paul Erdős, who famously said of explaining mathematical beauty: "It's like asking why Beethoven's Ninth Symphony is beautiful. If you don't see why, someone can't tell you."[4]

I will try, but in contrast to other expositions of mathematical beauty, I want to focus on the *experience* of beauty. How does order or clarity or elegance make you *feel?*

I can think of four types of mathematical beauty.

The first and most accessible kind of mathematical beauty is *sensory beauty.* This is the beauty of patterned objects that you experience with the senses: sight, touch, sound. Such objects can be natural, artificial, or virtual. Striking patterns in ripples of sand, the fractal pattern of a Romanesco cauliflower, and the stripes of a zebra are all produced by mathematical laws. Music is a pattern of sound waves that produces feelings of sensory beauty. Artwork in every culture includes patterns, sometimes created by using complex mathematical ideas. Quilting patterns are popular across the globe. Islamic art is particularly known for its intricate geometric designs. The Mandelbrot set, a striking geometric object with similar beauty at all scales of magnification, captured the public imagination in the 1980s after personal computers became powerful enough to compute it as a screen saver.

The feeling of sensory mathematical beauty is like the feelings that nature evokes. You feel joy, similar to when walking through a beautiful forest. You begin to esteem the order and the patterns that you see. You become attentive to little details. Your soul quiets. If you've ever been to the Sainte-Chapelle in Paris and bathed in the dazzling patterns of sunlight streaming through the stained glass, you'll know what I'm talking about. Architecture and music, because of their immersive mathematical natures, can enhance feelings of sensory beauty. I have this feeling whenever I dwell on Frank Farris's artwork illustrating Steiner's porism using floral rosettes (see next page). When you experience sensory beauty in the form of symmetry, smooth

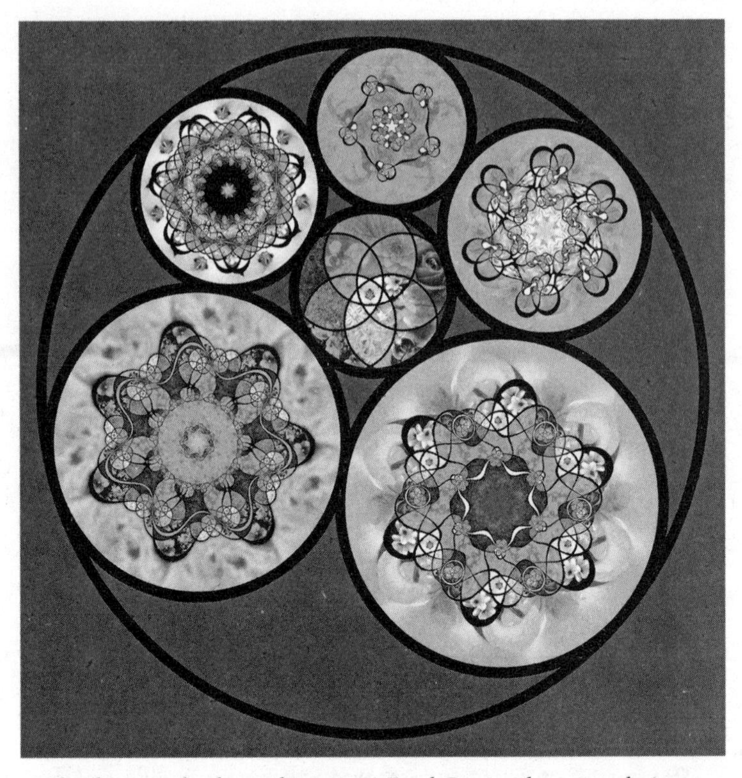

Floral rosettes by the mathematician Frank Farris, who uses techniques from complex analysis to create art from photographic material. This set illustrates Steiner's porism, a theorem about when circles can be inscribed in the area between two given circles in a way that completes a ring, like the one shown here. The outer circles are filled with rosette patterns whose colors come from the central floral collage (the original version is in color, but even in black and white this artwork is striking).

curves, and angles, aren't you just experiencing abstract algebra, calculus, and geometry in the wild?

This feeling explains why order and simplicity are often found in mathematical beauty. They evoke feelings of harmony and balance and a quietness of soul. So even though sensory beauty is the most accessible kind of mathematical beauty, it can be quite profound. You don't need to know any math to experience this kind of beauty. You appreciate it for what it is. Cherish sensory beauty, and it will lure you to perceive the world as a math explorer.

A second kind of mathematical beauty is what I'll call *wondrous beauty.* I use the word *wonder* in two senses: in the sense of feeling awe, at seeing something amazing, but also in the sense of feeling curiosity—that causes the mind to wonder, to be curious, to ask questions. While sensory beauty usually concerns itself with physical objects, wondrous mathematical beauty always invites a dialogue with *ideas.*

Wondrous beauty can follow from sensory beauty. If you look at a beautiful geometric design, you might begin to ask, How was it made? If you listen to an uplifting harmony, you may begin to ask, Why does it sound so glorious? Both *how* and *why* are the start of conversations with mathematical ideas. When you look at the Farris artwork, it causes you to speculate about how its striking patterns were generated. You don't need to have answers to your questions to experience wondrous beauty.

But wondrous beauty can be independent of sensory beauty. When a mathematician sees beauty in an equation, like $E = mc^2$, she isn't admiring it for its written physical form. She is admiring it for the idea contained within, which says that somehow energy and mass are interconvertible, with a small amount of mass equivalent to a large amount of energy. If she thinks this formula is beautiful:

$$e^{\pi i} + 1 = 0,$$

it is probably because there is no obvious reason why five of the most important constants in the universe should appear in the same equation. This is the surprise, the unexpected, that Hardy referred to as an element of mathematical beauty. It leads to wondrous beauty because surprising things cause us to become curious and to ask why.

M. C. Escher is a well-known Dutch graphic artist whose appeal is found in wondrous beauty. It's hard to walk away from his art without thinking more about it. His work often contains mathematical themes, such as symmetry or the infinite, while exploring the ambiguity of shifting frames of reference. He was fond of drawing impossible scenes, as in *Relativity* (1953) and *Waterfall* (1961), which have the curious property of being locally possible (in a small region, everything looks fine) but globally impossible (as a whole, the scene cannot be realized in the real world). His artwork is a prime example of wondrous beauty that invites a dialogue between a viewer and mathematical ideas.

In 1963, the mathematician Stanislaw Ulam was in a dull talk at a scientific meeting. Martin Gardner describes what happened next:

> To pass the time [Ulam] doodled a grid of horizontal and vertical lines on a sheet of paper. His first impulse was to compose some chess problems, then he changed his mind and began to number the intersections, starting near the center with 1 and moving out in a counterclockwise spiral. With no special end in view, he began circling all the prime numbers. To his surprise the primes seemed to have an uncanny tendency to crowd into straight lines.[5]

This phenomenon has been observed to hold for very large spirals (try it!), yet as I write this, the pattern remains without a satisfactory explanation. Ulam was experiencing the wondrous beauty that sometimes arises from play. When you see an unex-

100	99	98	97	96	95	94	93	92	91
65	64	63	62	61	60	59	58	57	90
66	37	36	35	34	33	32	31	56	89
67	38	17	16	15	14	13	30	55	88
68	39	18	5	4	3	12	29	54	87
69	40	19	6	1	2	11	28	53	86
70	41	20	7	8	9	10	27	52	85
71	42	21	22	23	24	25	26	51	84
72	43	44	45	46	47	48	49	50	83
73	74	75	76	77	78	79	80	81	82

The Ulam spiral. Primes (circled) seem to crowd in diagonal straight lines.

pected pattern like this, you cannot help but ask why. You feel both awe and curiosity.

A third type of mathematical beauty may be described as *insightful beauty.* This is the beauty of understanding. It is distinguished from sensory beauty, which concerns itself with objects, and wondrous beauty, which concerns itself with ideas. Insightful beauty concerns itself with *reasoning.* Logically correct deductions are not sufficient for the math explorer. She often looks for the best proofs, the simplest or most insightful. Math explorers have a special word for this: *elegance.* Paul Erdős often spoke of "The Book" that God keeps, in which all the most elegant proofs of theorems are recorded.[6]

Insightful beauty relies on elegant reasoning in the same way that the beauty of a poem relies on the words chosen. So insightful beauty has the surprising feature that it depends strongly on communication. What should be an elegant, lucid proof may be neither lucid nor elegant if communicated poorly. But when communicated well, it can move the soul like a poem, or induce delight like the surprise ending of well-told joke. Math explorers crave the feelings that insightful beauty generates.

The Sydney Opera House is one of the most iconic buildings in the world because of its architecture: the shell-shaped sections of its roof combine with the harbor setting to evoke the sails of a boat. But the story of how the roof took shape is a story of insight. The opera house's design is the result of a competition, and the winning entry by the architect Jørn Utzon, announced in 1957, was initially vague about the shape of the shells. Utzon's subsequent plan, for different parabolic shells, was unworkable from an engineering perspective. Owing to political pressures, construction began in 1958, before the cost and design of the roof were worked out. The roof design underwent more iterations, but costs were problematic, since so many different molds needed to be created for its sections and its tiles. And determining the border curves where the sections of the roof were to meet was a nontrivial mathematical challenge. Finally, in late 1961, Utzon had an epiphany. The Opera House's website describes what happened:

> Utzon was stacking the shells of the large model to make space when he noticed how similar the shapes appeared to be. Previously, each shell had seemed distinct from the others. But now it struck him that as they were so similar, each could perhaps be derived from a single, constant form, such as the [surface] of a sphere.
>
> The simplicity and ease of repetition was immediately appealing.
>
> It would mean that the building's form could be pre-

The Sydney Opera House.

> fabricated from a repetitive geometry. Not only that, but a uniform pattern could also be achieved for tiling the exterior surface. It would become the single, unifying discovery that allowed for the distinctive characteristics of Sydney Opera House to be finally realised, from the vaulted arches and timeless, sail-like silhouette of the Opera House to the exceptionally beautiful finish of the tiles. . . .
>
> By any standard it was a beautiful solution to crucial problems: it elevated the architecture beyond a mere style—in this case that of shells—into a more permanent idea, one inherent in the universal geometry of the sphere.[7]

The flash of insight that Utzon had is often described as an *aha!* moment. In mathematics, it's that thrill of sudden understanding, when something foggy becomes crystal clear—like finding an elegant solution or illuminating proof. The emotion that follows is the giddy excitement often associated with seeing the big picture, when everything finally makes sense.

Insightful mathematical beauty is like the feeling, while you're shopping, of stumbling on a sale for an item you never knew you needed that meets desires you never knew you had. It's like the feeling of watching the ending of a mystery movie, when everything gets explained. And just as people watch movies a second time to see new details, a math explorer enjoys dwelling a second time on the threads of an insightful argument, to think about its ramifications, generalizations, or applications.

Now, it's true that clumsy solutions can also be found suddenly, but the associated emotion in that case is not excitement, but relief—that the ordeal is over. Long-winded arguments tend to be clumsy and are easily forgotten; it's no wonder that the simplicity and lucidity of an argument are associated with beauty.

Puzzles with insightful beauty often get shared. On the facing page is one I heard from a random person at a math conference.

There's an *aha!* moment to be had when you see a very simple and elegant way of looking at this problem. (For a hint, see Hints & Solutions to Puzzles.)

Insightful beauty can manifest itself in a flash of insight, or in a slowly growing appreciation over time. There are many mathematical ideas that I didn't appreciate until I had seen them arise over and over again in disparate places. One recurring theme throughout mathematics is *duality,* natural pairings that exist between mathematical ideas: examples include multiplication and division, sine and cosine, unions and intersections, points and lines. Recognizing duality is like using a mirror to see how two creatures that look and behave differently are really the

ANTS ON A LOG

One hundred ants are dropped at random places on a log, each ant facing one end or the other. The log is 1 meter long, extending left to right. Each ant travels toward either the left or the right end with constant speed 1 meter per minute. When two ants meet, they bounce off each other and reverse direction, keeping their speed intact. When an ant reaches an end of the log, it falls off. At some point all the ants will have fallen off. Question: Over *all* the possible initial configurations, what is the longest amount of time you would need to wait to guarantee that the log had no more ants?

same. I didn't appreciate duality until I saw it in many contexts; now I think it is beautiful.

The deepest experience of mathematical beauty is to be found in *transcendent beauty.* It can amplify or be amplified by sensory, wondrous, or insightful beauty, but it goes well beyond. Transcendent beauty generally arises when one moves from the beauty of a specific object, idea, or reasoning to a greater truth of some kind—perhaps an insight that reveals its deep significance, or a deep connection to other known ideas. When you experience this kind of beauty, you feel a profound awe and even a sense of gratefulness. The mathematician Jordan Ellenberg, in his book *How Not to Be Wrong,* describes it this way:

> What's true is that the sensation of mathematical understanding—of suddenly knowing what's going on, with total certainty, *all the way to the bottom*—is a special thing, attainable in few if any other places in life. You feel you've reached into the universe's guts and put your hand on the

wire. It's hard to describe to people who haven't experienced it.[8]

Many math explorers have felt this transcendence by asking this metaphysical question: Why should mathematics be as powerful as it is to explain the world? Albert Einstein asked, "How can it be that mathematics, being after all a product of human thought which is independent of experience, is so admirably appropriate to the objects of reality? Is human reason, then, without experience, merely by taking thought, able to fathom the properties of real things?"[9] He was expressing awe at experiencing transcendent beauty, and in embracing his wonder, we experience it too.

Math explorers also find transcendence in theories that unify disparate areas, and sometimes the language reflects it. "Monstrous moonshine" was the fanciful name given to an unexpected link discovered in the late 1970s between number theory and a colossal symmetry structure known as the monster group.[10] Strangely, the coefficients that showed up in an important function in number theory also appeared as sums of important dimensions of the monster group. In 1992, Richard Borcherds proved that the conjectured connection was really there by showing, astonishingly, that both are connected to string theory! This work eventually earned him a Fields Medal, and in an interview afterward he said that winning the honor was not as exciting as solving the problem. He described his feelings: "I was over the moon when I proved the moonshine conjecture. If I get a good result I spend several days feeling really happy about it. I sometimes wonder if this is the feeling you get when you take certain drugs. I don't actually know, as I have not tested this theory of mine."[11]

Transcendent mathematical beauty is not either off or on. It can come in degrees, proceeding from sensory, wondrous, or insightful beauty. We may feel it when the sensory geometric beauty of a majestic architectural space hits us at a deep level. Or when we see a simple idea appear in many different forms across multiple areas of mathematics. Or when we grasp that a certain elegant proof can generalize to many other situations.

The transcendent beauty we find in the world provokes the feeling that there's something beyond our grasp, waiting to be found, that may be of ultimate meaning. C. S. Lewis spoke of sublime experiences of beauty as "the scent of a flower we have not found, the echo of a tune we have not heard, news from a country we have never visited."[12] In a similar way, mathematics can feel transcendent. When you see the same beautiful idea pop up everywhere, you begin to think that it is pointing to some deeper truth you haven't yet grasped. When you realize that you've had exactly the same mathematical thoughts as another person—separated from you by oceans, culture, and time—you begin to believe there might be a universal, enduring reality that you are both somehow accessing. There are whispers calling us, but we have not yet found their source.

Pursuing beauty of any kind cultivates in us the virtues of *reflection, joyful gratitude,* and *transcendent awe.* When I took a five-day hike through the High Sierras, I had the opportunity and time to experience the beauty of the wilderness. Crunching through an icy meadow in the middle of July, witnessing its shimmering splendor and reflecting on the fact that I might have been walking where no one had walked before, I felt a sense of gratitude and awe. The pursuit of mathematical beauty offers some unique ways to cultivate these virtues, while motivating math explorers to study mathematics. As with my hike in the Sierras, chasing after mathematical beauty may lead you to wonderful spaces no eye has yet beheld, allowing you to experience a profound idea in a totally new way. And taking time to reflect on beauty makes us better equipped to learn math and process new information. In this digital age, when we are bombarded with notifications and other distractions, we need space for reflection more than ever.

Appreciating transcendent mathematical beauty builds in us *habits of generalization,* when we look for overarching patterns where we might not expect them. When I learn a new theorem, I often ask, What gives this theorem its power? What is the underlying principle? How might it apply to a more general situation? Such habits carry over to other areas of my life. When I'm cooking a stir-fry from a new recipe, I often ask: What general principles does this recipe teach me? *Add garlic and chopped onions first. Put the basil in at the end, or it will lose its color.* This habit of looking for general cooking principles allows you to improvise new delights.

If mathematics is for human flourishing, then we can all benefit from grasping its beauty. But there are many notions of beauty, and many ways to motivate our study of mathematics through beauty—through art, through music, through patterns, through cultural artifacts, through rigorous arguments, through the elegance of simple but profound ideas, through the wondrous applicability of these ideas to the real world in many different fields.

To make these connections for others, you must appreciate what they find beautiful in life. Is it sensory, wondrous, or insightful beauty? Hearing their stories will give you a way to connect their notions of beauty to how they could experience mathematics.

Unfortunately, math can be taught in a way that stifles its beauty. Learning math as a bunch of rules without meaning-making insights or as an endless stream of repetitive problems that lack joyful resolution is a surefire way to sap desire. A recent op-ed in a major newspaper told people to "make your daughter practice math. She'll thank you later."[13] The op-ed never asked the question of how to teach math so she'll thank you *now.* The beauty of math can inspire her to practice if you give her delightful problems with unexpectedly elegant solutions—these will transform a dull practice into an exciting exploration. Those who have tasted beauty and its virtues will welcome practice as a way to taste it again and again.

For when we experience beauty that stirs us, we long for more. Of all the virtues cultivated by mathematical beauty, this may be the most important one of all: the *disposition toward beauty.* It's like how reading a good book by an author makes you want to read another book by the same author. It's like how learning a new word makes you want to use that word as much as possible. It's like how the joy you feel when you exercise makes you want to exercise every day.

A disposition toward mathematical beauty is the engine of mathematical persistence. No matter how hard the problems get, you'll keep coming back—because you know: every new mathematical challenge brings the hope of beholding beauty once more.

CHESSBOARD PROBLEMS

Here's a classical problem you may enjoy thinking about, with what many people consider to be a beautifully elegant solution.

Think of a chessboard, which is an 8×8 grid of squares. Suppose you have a bunch of 1×2 dominoes, each of which can cover two adjacent squares of the chessboard. We say the dominoes will "tile" the chessboard, because you can cover the whole board with dominoes that do not overlap and do not overhang the board.

Suppose you remove two corner squares that are diagonally opposite each other on the chessboard. Can you tile the chessboard (minus those two corner squares) with dominoes? If so, exhibit a tiling, and if not, show that a tiling is impossible.

Christopher solves this problem in the next letter, so if you don't want to see the solution, skip his first paragraph. But now here are more questions you can explore:

- Consider other chessboards that have two squares removed. For which ones is it possible to tile the board with dominoes?
- Place a knight on each square of a 7×7 chessboard. Is it possible for each knight to simultaneously make a legal move?

(A legal move for a knight is to move to a square that is 2 squares away in one dimension and 1 square away in the other.)

- Can you cover a 4×7 chessboard using all seven distinct Tetris pieces (the seven possible quadrominoes—each with 4 squares) at the same time?
- Consider an 8×8×8 cube, composed of 512 little cubes. Remove two cubes from opposite corners. Can you tile what's left with 1×1×3 blocks?

February 2, 2018

Hello, Professor Su, I hope you had a better week than the last one. I believe I have an answer to the puzzle. I believe the answer is that it is impossible to tile the modified chessboard. Here's why: In order to tile the whole chessboard in any way, each domino has to cover a white square and a black square. Every diagonal opposite corner square is of the same color. Therefore you will have either 2 extra black squares or 2 extra white squares. Same-color squares are always diagonal and therefore cannot be covered by a domino. You will always have 30 (black or white) squares trying to be covered with 32 (white or black) squares, so 2 (white or black squares) will be uncovered. I hope that this is sufficient. I hate to have an answer where I could say it's impossible, because it feels like I'm somehow tapping out (quitting).

Since you say that linear algebra should be studied early I'll study that subject next. I remember reading an excerpt about a book describing reality in non-linear equations and it referred to Einstein's work. I also remember reading that the Greeks knew that Euclidean geometry wasn't necessarily an exactly accurate portrayal of reality. I'll try to go on and get to my linear algebra books next: it's 3 of them and one of them is 505 pages and 2,400 proof problems, but I'll get to it.

Topology is a very interesting subject, rather abstract. The topology I'm studying right now is, I guess, introductory (Point-Set; he also called it a general topology in the preface), so it hasn't really gotten too much into deformations. The main subjects have been so far topological spaces, axioms of separation, compactification and uniformization, and now I'm in the chapter titled "Continuity." . . .

How do you know when you've written a good proof or a good demonstration? Also I was back in my rigorous calculus book and I wrote a proof explaining why a cube in R^3 is a normal domain. And I think I wrote a way better proof than I was writing before because I've been doing topology proofs. Do you think proofing in certain subjects is easier than in others? I so far think that the areas of mathematics that interest me are analysis, number theory, and computation theory. I have a couple of books touching on computation and 1 analysis book; I don't have a number theory text yet, but from what I've read on it, it definitely seems interesting.

Chris

6
permanence

With every simple act of thinking, something permanent, substantial, enters our soul.
Bernhard Riemann

I felt as though I had been granted access to a world which was, until then, completely invisible to me. I was—and still am—enamored [of] how mathematics is so expertly woven into the world around us.
Tai-Danae Bradley

I have an old flannel shirt in my wardrobe. Its leafy hues remind me of some of my favorite hikes through the forest. Its versatility means that this shirt was convenient to grab and wear

on many occasions—a walk in the mountains, gatherings with good friends, late-night rap sessions with dorm buddies chatting about the meaning of life—and just like a security blanket, its warm woolen fibers were a consolation to me. Now it's fraying, showing signs of age, but I have so many memories associated with it that I refuse to throw it away. Its raggedness and ruggedness speak to me, telling stories of yore. We all long to have fixtures in our lives, things we can depend on, and that shirt has been with me through thick and thin, and I will not abandon it.

My trusty flannel shirt represents permanence.

Permanence is a universal human desire. We hope that beauty and love are *eternal.* We seek *immortality* or at least try to put off death as long as possible. Don't we extol the virtues of *endurance?* Don't we pledge our love *forever?* We speak of leaving a lasting *legacy,* or making our *mark,* in hopes that the mark (if not ourselves) will be permanent. And the desire to have children is in some ways a longing for permanence.

Permanence is also a mathematical longing. Consider the attention that math explorers give to unchanging things.

We adore *constants.* The word *constant* is often used to describe some fixed number that has significance, like the golden ratio, e, or π. The number π captures the public imagination in part because it arises in many places, not just in the geometry of circles. Some instances are quite unexpected, such as in the sum of reciprocals of square numbers—

$$1/1^2 + 1/2^2 + 1/3^2 + \ldots = \pi^2/6$$

—or in the formula for the area under a bell curve, or in Heisenberg's uncertainty principle. As such, the number π feels cosmic—mysterious and eternal, no doubt an important constant at any point in time and in any corner of the universe. It is no surprise that some people try to grasp this number's inscrutable nature by memorizing its digits or seeking patterns in them.

We also search for *invariants.* In mathematics, an invariant is something that remains unchanged when you perform an operation. For instance, if I multiply a number by 5, that doesn't change whether it is even or odd. If I rotate a 3-D geometric figure, that won't change its volume. Invariants reveal insights about the operation itself. For example, I can learn something about a rotation by what it doesn't change: the axis about which the rotation spins is fixed and reveals an important feature of the rotation. A key to solving a Rubik's cube is paying attention to what doesn't move when you make a move. In mathematical models of physical systems, a conservation law focuses on an invariant quantity that doesn't change as the system evolves—such as the total momentum when two cars collide. Invariants are fixtures that we begin to recognize and rely on.

Invariants can tell us when something is impossible. For instance, at the end of the last chapter, I left you with this puzzle: using dominoes, can you tile a chessboard with a pair of opposite corner squares removed? This classic problem is made easier by looking for an invariant: something that doesn't change when you place a domino. (Spoiler alert: skip to the next paragraph if you don't want to see the solution.) Since a domino always covers one black square and one white square, no matter how many dominoes you place, the number of covered black squares always equals the number of covered white squares. This equality is an invariant, and it gives us the following insight: if you remove *any* pair of squares of the same color from the chessboard (such as opposite corner squares), the number of white and black squares will be unequal and you will not be able to completely cover this board.

The names of many other mathematical concepts also reflect an interest in permanence: *stable sets, convergence, equilibria, limits, fixed points.*

Mathematical ideas themselves have a permanence that is attractive and beautiful to us. The physical sciences speak of nature's *laws,* which are usually facts that are empirically observed to hold true over many instances.[1] These can sometimes be overturned by new knowledge. But in mathematics, we speak of *theorems.* When proof establishes a theorem, it will never be overturned. It is true for all time, and it is true everywhere in the universe.

This kind of permanence is unique to mathematics. David Eugene Smith, in his 1921 presidential address to the Mathematical Association of America, spoke of the immortality of mathematical laws:

> The laws of the Medes and Persians, unchangeable though they were thought to be, have all perished; the canons that bound Egyptian activities for thousands of years exist only in the ancient records, preserved in our museums of antiquity; the laws of Rome, which at one time dominated the legal world, have given place to modern codes; and the laws that we make today are certain to be changed tomorrow. But in the midst of all these changes it has ever been true, it is true today, it shall be true in all the future of this earth, and it is equally true throughout the universe whether in the algebra of Flatland or in that of the space in which we live, that $(a + b)^2 = a^2 + 2ab + b^2$. . . .
>
> What I learned in chemistry, as a boy, seemed true at the time, but much of it today is known to be false. What I learned of molecular physics seems at the present time like children's stories, interesting but puerile. What we learn

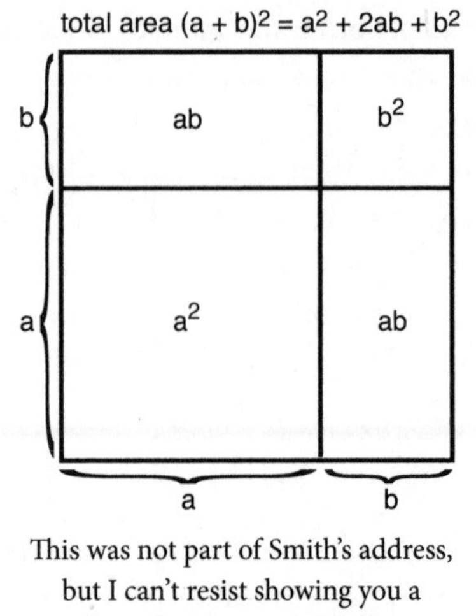

This was not part of Smith's address, but I can't resist showing you a "proof without words" that $(a + b)^2 = a^2 + 2ab + b^2$.

in history may be true in some degree, but is certain to be false in many particulars. So we may run the gamut of learning, and nowhere, save in mathematics alone, do we find that which stands as a tangible symbol of the immortality of law, true "yesterday, today, and forever."[2]

Why do we seek permanence?

We seek permanence because it is a *refuge,* and stability is a safe haven. I grab my flannel shirt because it is familiar and makes me feel at home. I make promises, as I have done in marriage, not because keeping them will be easy but because keeping them will be hard. Then my spouse has the safety to rely on those promises, and I find safety in hers. In the ups and downs of life, permanence is comforting.

The permanence of mathematics can be a refuge as well. We can find joy in a timeless puzzle that engrosses us so completely that we tune out our troubles. Engaging the mind in creative problem solving can lift our spirits and give us another way to be. The math popularizer Morris Kline, in his book *Mathematics for the Nonmathematician,* expresses this kind of rest and comfort:

> The tantalizing and compelling pursuit of mathematical problems offers mental absorption, peace of mind amid endless challenges, repose in activity, battle without conflict, and the beauty which the ageless mountains present to senses tried by the kaleidoscopic rush of events.[3]

After the 1941 bombing of Pearl Harbor, some 120,000 Japanese Americans in the United States were forcibly stripped of their homes and possessions and relocated to internment camps. With no furniture in these bleak settings, they had to fashion substitutes from scrap lumber and metal. So carving, sculpting, painting, designing, and creating arts and crafts became a way to practice *gaman*—a Japanese word meaning "to endure the unbearable with dignity and patience."

The exhibit *The Art of Gaman* showcases extraordinary pieces of art fashioned by ordinary men and women in these camps.[4] Among the artwork, you'll find a hand-painted wooden sliding block puzzle that Kametaro Matsumoto made for his children. The object of this geometric puzzle (see next page) is to slide the blocks so that the young woman (large square tile) can escape her parents and their hired workers (domino tiles) and exit the rectangle (at bottom middle), with four eager suitors (small square tiles) close behind. This puzzle appears to be a variant of one that was popular in the 1930s, also known as *Hakoiri musume* (Japan), *L'âne rouge* (France), and *Hua rong dao* (China). Each version has different characters. The fewest number of moves is known to be eighty-one steps.[5]

Sliding block puzzle, made by Kametaro Matsumoto, in its starting configuration (left) and in an ending configuration (right). Photos courtesy of Shinya Ichikawa; sliding block puzzle courtesy of Jean Matsumoto and Alice Ando.

Making the puzzle and solving the puzzle were both ways in which mathematical thinking was a refuge, and a vehicle for gaman. This is a true picture of what it looks like to flourish in even the most difficult circumstances.

A second reason we seek permanence is that it is a *yardstick* against which we measure our progression through life. My flannel shirt now has imperfections, small stains and tears that mark certain events and memories. It's not the same shirt that it was before. And I'm not the same person either. It was a loose-fitting shirt when I was in my twenties; now it's tighter—in certain places. So each time I grab that shirt, I look at it differently. It fits differently. It has new meaning for me each time, because I can mark my progression through life by that shirt.

Similarly, while the truth of a theorem is eternal and unchanging, every time I come to it I see it differently. I first come to it in the struggle of trying to grasp this truth, and it is an angry bear. The next time I return to it, I may experience the thrill of understanding, and it is a tamed lion. Then years later I return to it and I see a friendly dog, who's chummy with all the other pets in the neighborhood—the other truths that I know. The theorem is an ever-present yardstick against which I can measure my progress, reminding me of my struggles and victories. Perhaps you can think of mathematical ideas you found hard to grasp as a child that are second nature to you now and feel thoroughly familiar. For math explorers, each theorem is a nostalgic souvenir of a past adventure that stretched one's capabilities.

A third reason we seek permanence is that it is a *foothold* on which we can rely. When I climb a rocky wall, I must know where to set my feet. Of course, I don't look for sandy crevices—I look for rocks that are immovable. In the same way, when I visit an unfamiliar place, I seek out landmarks, unchanging features by which I can navigate, just as for centuries humanity has navigated by the stars, the fixtures of the firmament. And when I reach for my flannel shirt, I see it as a reliable fixture of my wardrobe, a dependable garment around which I'll build the rest of my outfit.

In mathematics, axioms, definitions, and theorems often serve as footholds when we try to prove other statements. The Greek mathematician Euclid is famous for his treatise *The Elements* (written around 300 BC), which systematically organizes geometric results by logically deriving them from certain axioms and postulates considered self-evident. But the Euclidean axioms are not the only way to organize geometry. The mathematician David Hilbert, in 1899, chose a different set of axioms.

There are many ways to climb a rocky cliff; Euclid and Hilbert chose different footholds as starting points. Axiomatic systems have been developed for other fields of mathematics too. Rarely does the everyday math explorer need to resort to using such axioms, because we're already scaling the cliff, moving from rock to rock, but it can be comforting to know how to trace a path from the ground that will get us up here.

And a new mathematical theory often begins with initial postulates or definitions as starting points for exploration. Einstein's theory of special relativity, first published in 1905, begins with two starting postulates: the laws of physics and the speed of light are the same for all observers who are not accelerating. These invariance assumptions led Einstein to deduce that length, mass, and time *do* depend on one's frame of reference, very strange mathematical conclusions that have since been experimentally confirmed.

Whereas axioms are like footholds near the ground, theorems are more like footholds farther up the cliff. The especially useful ones are like large ledges from which you can push off in multiple directions, or on which you can just rest and enjoy the view. Theorems summarize key findings and provide foundations on which to build applications. The central limit theorem is a probabilistic result that explains a striking phenomenon: when you take a sufficiently large random sample from a population to measure a certain quantity in that population (e.g., the percentage who found a drug effective), the sample average of that quantity will have a bell-shaped distribution, regardless of the unknown population distribution of that quantity. The central limit theorem is the foundation for many statistical applications, such as computing confidence intervals for a population average or deciding whether a test result is strong enough to make a conclusion (e.g., a certain drug is more effective than a placebo).

So we seek permanence because it is a refuge, a yardstick, and a foothold. But that does not fully capture why this is such a deeply embedded human longing.

Every human longing contains at its core a question of ultimate significance. If you desire to love and be loved, you have to wrestle with the question "Am I lovable?" If you have a desire for beauty, you must ask, "What is good?" If you have a desire to play, your soul is acknowledging a deep idea: that there is more to life than work. And the longing for permanence has, at its core, an unmet need.

Whom or what can I trust?

Trust is at the heart of a desire for permanence. If I seek a refuge, it is because I need a hiding place that I trust to be safe. If I look to a yardstick, it is because I expect it will not change. If I step in a foothold, I must have confidence that it will hold me up.

As I write this, the United States is a deeply divided nation in a crisis of trust. People are asking, Can I trust someone who has an utterly different worldview than mine? Can I trust our political leaders? Can I trust the media? We have family members who can't tell what news is fake and what is real, and in the shifting shadows, some have given up trying. Partisan bickering has resulted in a lack of intellectual safety, because there is no firm foundation to what people believe they can know. So they just give up. George Orwell, in his dystopian novel *Nineteen Eighty-Four,* describes a world in which the totalitarian government (the Party) manipulates the people through propaganda and deception:

> In the end the Party would announce that two and two made five, and you would have to believe it. It was inevitable that they should make that claim sooner or later: the logic of their position demanded it. Not merely the validity of experience, but the very existence of external reality, was tacitly denied by their philosophy. The heresy of heresies was common sense. And what was terrifying was not that they would kill you for thinking otherwise, but that they might be right.[6]

What the permanence of mathematics offers is trust that mathematical reasoning is solid ground that will not move. *Trust in reason* is a virtue built by acknowledging the permanence of mathematics. The arguments that worked yesterday will still work today. They establish unshakeable truths that can be known through deep investigation and reason. Thus, it is no surprise that Orwell chose a mathematical falsehood for what the Party would force you to believe. Most of our knowledge in life involves points of view, uncertainty, error, and incomplete information—so our knowledge is subject to revision. But mathematical truths don't get overturned. Their interpretations may change or they might become irrelevant, but their truth stays the same: yesterday, today, and forever. For Orwell, the most terrifying absurdity one can imagine is the loss of our footholds, our yardsticks, and our refuge when mathematical truths—the very existence of external reality—are no longer permanent.

SHOELACE CLOCK

You are given a shoelace, some matches, and a pair of scissors. The shoelace burns like a fuse when lit at either end and takes exactly 60 minutes to burn. The burn rate may vary from one point on the shoelace to another, but it has a symmetry property in that the burn rate a distance x from the left end is the same as the burn rate the same distance x from the right.

1. Find the shortest time interval you can measure. (For example, 30 minutes can be measured by lighting each end of the shoelace and waiting until the flames meet.)
2. Find the shortest time interval you can measure if you have two such laces that are identical.[a]

a. Puzzle proposed by Richard I. Hess in "Problem Department," ed. Clayton W. Dodge, *Pi Mu Epsilon Journal* 10, no. 10 (Spring 1999): 836. Hess credits Carl Morris for the original idea.

September 9, 2018

Do I find mathematics a refuge? As I continue in my study of mathematics, mathematics is training me. Tenacity, patience, humility, confidence in reason to find an answer, the imperative to build reasoning, things which I had and believed in before my serious study of math, but I find that as I continue, these qualities are being perceptibly strengthened by dealing with math. Math has been a refuge for me since I've been serious about it my whole time I'm in prison.

When I open *God Created the Integers*[a] and read and understand Euler's theorem that every integer is the sum of four squares or study Fourier's "Analytical Theory of Heat" and understand it, I know I'm grappling with ideas that were or are toward the pinnacle of human understanding and are eternal. As long as this universe exists, 2 + 2 will always equal 4, the number of degrees inside a triangle will always be . . . whatever it is, once we figure it out. It makes me feel closer to something larger and powerful and deeper. Closer to . . . maybe truth.

Chris

a. *God Created the Integers* (Philadelphia: Running Press, 2005) is a book by Stephen Hawking containing excerpts from important papers in the history of mathematics. This is one of Chris's favorite math books.

7
truth

What is truth?
Pontius Pilate

Truth is so obscure in these times, and falsehood so established, that unless we love the truth, we cannot know it.
Blaise Pascal

"Our teacher told us you were adopted. I didn't know you were adopted."

I didn't know either. I was in sixth grade when my friend informed me, and I suspected he was right. Through the small-town grapevine, people find things out about you that you don't

know about yourself. Until that point, I had only a suspicion. I didn't think I looked anything like my family. We didn't have any baby pictures at home. There were the occasional slipups: I still remember when a guest said, "You've grown so big! I remember when your parents first got you," and then an *oops* look appeared on her face.

My friend's declaration had a ring of truth because it explained many things that didn't make sense before. I could have ignored it and continued as though nothing were different, just like I had so many times in the past when I had only hunches. But somehow hearing it said aloud popped my bubble of denial. I had to find the truth.

Truth is a basic human desire. We crave it, even if it might bring us uncomfortable information. Sometimes we don't act on that desire, but it gnaws at us. After I confirmed my adoption, I knew I wanted to find my biological family, even if it meant learning some hard truths about my past. Yet I waited for many years to do anything about it. There are many reasons we don't act on our desire for truth. When life gets hectic or unmanageable, we tell ourselves, "I can't deal with this right now," and we choose to live in a bubble—but it'll pop someday. Will we be prepared when it does?

The world I see today is being upended by political instability, stoked by misinformation. Truth is obscure. People are endorsing blatant falsehoods that comport with their worldview rather than accepting complicated truths. We live in filter bubbles that reflect our own biases back at us. Falsehoods are routinely shared within these bubbles, sometimes maliciously. And ironically, the groups responsible for wielding lies as weapons have been unwittingly assisted by opposing groups who have, until now, questioned whether objective truth exists. Pilate's rhetorical question "What is truth?" embodies the exasperation of a confused public. I hear the same hopelessness when my friends lament, "It's just too hard to figure out what's going on, so why should I even bother?"

Why should anyone bother? Does truth matter anymore? Should we just blindly trust authority to decide what's true? Or can we discern truth for ourselves?

Truth is a mark of human flourishing. A flourishing society values truth, while oppressive societies suppress it. Think of totalitarian regimes, which control their media and propagandize their own people. The political theorist Hannah Arendt studied such regimes. In her 1967 article "Truth and Politics," she says:

> The result of a consistent and total substitution of lies for factual truth is not that the lies will now be accepted as truth, and the truth be defamed as lies, but that the sense by which we take our bearings in the real world . . . is being destroyed.[1]

When we lose our bearings, we begin to care less about what's true. We become more easily manipulated. *Why should I even bother?*

Mathematical thinking equips us to figure out what's going on, and to bother. Math explorers care about deep knowledge and deep investigation.

I've been using the word *truth* and realize that we may have different ideas about what that word means. I take it in the sense that most people do: true statements are ones that align with reality.[2] This definition sweeps under the rug a whole host of philosophical questions—like "What is 'reality'?"—but the ways in which I will discuss truth won't require more nuance. If I say, "The sky is blue," that is a statement about the physical world that is easily verified. This truth may admit some subjectivity, depending on what *blue* means to the observer, but there is a sense in which the statement aligns with reality. If pressed, we can define our terms and measure the wavelengths of emitted light. If I say, "I was adopted," that is a historical statement. While it cannot be verified in the same way that physical truths can, significant evidence points to its veracity—I now have multiple ways of knowing it. In any case, there is a historical reality in which I either was or was not adopted. That reality establishes the truth, no matter what my belief is.

In no way do I deny the complexities of understanding truth or the subjective lenses through which we interpret the world—if anything, I'm arguing that we embrace complexity. When I got in touch with my biological family, I had to sort out this uncomfortable question: Why was I given up for adoption? I asked many people. Every time someone said, "Your biological mother gave you up for adoption because ______," I had to look at that statement through multiple lenses: their lens (why were they saying that?), my lens (how did this make me feel?), and my biological mother's lens (what would she say?). Such statements were complex, and open to subjective interpretation. Nonetheless, taken together, the answers produced an emerging picture of a whole truth that I would not have gotten if I hadn't been willing to embrace the complexity of whatever truth I might find.

For a math explorer, *deep knowledge* of truth is essential. The claim of truth is not the same as a deep understanding of truth. When an explorer does a calculation, like 777 × 1,144, she isn't satisfied by just getting an answer: 111,888. She deeply understands what multiplication means, so she can see if the answer is reasonable, or if she punched something wrong into her calculator. She knows that the last digit of the answer (8) is determined only by the product of the last digits of the numbers being multiplied ($7 \times 4 = 28$), and they match up in this case. She knows that since the first number is more than 700 and the second number is more than 1,000, her answer should be at least 700,000. It isn't, so she knows she made an error. Understanding deeply means that you can check whether your answer is reasonable. As math explorers, you and I will make math mistakes, just like everyone else. We're just more likely to catch our errors.[3] The claim of truth (in this case, a flawed calculation) is not the same as the deep understanding of truth (which offers multiple ways to check an answer).

Similarly, at higher levels of mathematical knowledge, the truth of a statement must be fathomed in a deeper way. As the mathematician Gian-Carlo Rota writes:

> What matters to any teacher of mathematics is the teaching of what mathematicians in their shoptalk informally refer to as the "truth" of a theory, a truth that has to do with the concordance of a statement with facts of the world, like the truth of any physical law. In the teaching of mathematics, the truth that is demanded by the students and provided by the teacher is such a factual/worldly truth, not the formal truth that one associates with the game of theorem-proving. A good teacher of mathematics is one who knows how to disclose the full light of such factual/worldly truth before students, while at the same time training them in the skills of carefully *recording* such truth.[4]

The game of theorem proving may lead you astray if you don't grasp the truth of the theorem first. Many times when I'm trying to solve a problem I don't understand, I will make a sequence of logical steps, one after the other, and end up with an erroneous statement. I've made an error somewhere and I don't see why, because I don't really see the truth of what I'm trying to prove. Experiencing this confusion multiple times makes me want to understand things deeply. Math explorers are not satisfied with shallow knowledge.

For a math explorer, *deep investigation* of truth is a habit. To really know truth is to investigate it at many different levels—to see the whole truth from multiple perspectives. Explorers seek multiple ways of knowing the truth, not just to check their work, but to see deeply how they are in concordance with one another. This can mean trying lots of examples to get an intuition about what's going on. It can mean running experiments, much as a scientist would, to gather evidence that proves or disproves a conjecture. It can mean proving a theorem in multiple ways. It's part of constructing meaning, making sense of the truth: do the stories match? For instance, the study of linear algebra concerns itself with, among other things, solving systems of linear equations, like these:

$$x + 2y + z = 8.$$
$$3x - y + 5z = 16.$$

When you ask for simultaneous solutions of both these equations, you are looking for numbers x, y, and z that satisfy both equations. For instance, $x = 1$, $y = 2$, $z = 3$ works. But there are many, many other solutions—they go on forever, an unlimited number. In mathematics we say the set of solutions is *infinite*. If you learn the technique called Gaussian elimination, you can verify that these two equations have infinitely many solutions.

But a math explorer would additionally seek out other ways of understanding this truth. She might rewrite these equations as a single equation:

$$x\,(1, 3) + y\,(2, -1) + z\,(1, 5) = (8, 16).$$

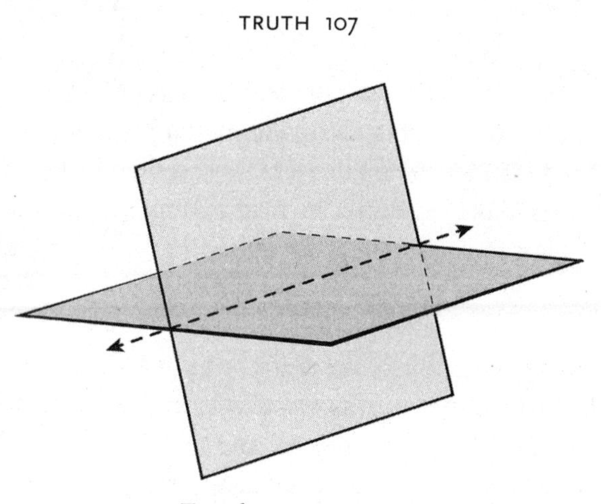

Two planes intersecting.

This is now a geometric question: if you have a spaceship at the origin (0, 0) in a two-dimensional plane, and it has three thrusters that can push you in the directions (1, 3), (2, −1), and (1, 5), will some combination allow you to reach (8, 16)? From that perspective, any two of these thrusters will suffice to move you about this two-dimensional plane, so the math explorer sees that there are multiple solutions, though it might be less apparent that there's an infinite number.

She might next think about the system differently, by recalling that linear equations like these have solution sets that are flat planes in three-dimensional space. Their intersection gives the set of solutions to both equations. But two flat planes that intersect must intersect in a straight line, so she sees that the set of solutions must be a line, and there are indeed infinitely many points on a line. She's now confirmed this truth in multiple ways. Deep investigation goes hand in hand with deep understanding.

Because deep investigation is a math explorer's habit, she will often take a truth deeply grounded in reality and extend it far into the imagination, to envision new realities. These realities may be Platonic, existing in an ideal form as purely intellectual concepts. It is an unexplained power of mathematics that these envisioned realities sometimes end up describing aspects of the physical world that we didn't see before. Who would have imagined that linear algebra ideas developed in the mid-nineteenth century would find spectacular application in the twentieth century in quantum mechanics or in the mathematics of search engines like Google? The Nobel Prize–winning physicist Eugene Wigner spoke of the "unreasonable effectiveness of mathematics" to explain the natural sciences, and said: "The miracle of the appropriateness of the language of mathematics to the formulation of the laws of physics is a wonderful gift which we neither understand nor deserve."[5]

The mathematician Georg Cantor (1845–1918) is famous for exploring the nature of infinite sets. A normal person would say that all infinite sets seem like they're the same size—they go on forever, and there's not much to say beyond that. But the math explorer asks, How can we make sense of the "size" of infinite sets? How do you count something that goes on forever? What Cantor realized is that you *can* "count" infinite sets, not by using numbers but by using other sets! You try to pair up each element of one set with an element of another set so that nothing's left unpaired in either set. If you succeed, the element-wise pairing is called a *one-to-one correspondence,* and we say the two sets have the same *cardinality*—that's the mathematician's word for "size."

The surprise of Cantor's work, which he first published in 1874, is that infinite sets come in many sizes—infinitely many, actually! Not only that, but it turns out that the set of whole numbers (like 0, 1, 2, 3, . . .) does *not* have the same cardinality as the set of all real numbers between 0 and 1 (think of real numbers as numbers whose decimal expansions are finite or go on forever without necessarily repeating). This was a major surprise, and many mathematicians initially refused to accept Cantor's theories. This crazy truth about infinite numbers seemed farfetched and disconnected from reality in 1874. But today we see that it implies some interesting things about the limits of computation. Since one can show that the set of all computer programs has the same cardinality as the set of whole numbers, this means that there are some real numbers whose decimal digits cannot be sequentially generated as the output of a computer program! Math explorers, habituated to deep investigation, often stumble on some surprising truths.

The quest for deep knowledge and deep investigation of truths in mathematics builds many virtues that carry over to other areas of life. The first is the *thirst for deep knowledge and deep investigation* of any truths of importance.

When shallow knowledge has led you astray too many times in mathematics, you begin to crave knowing things in a deeper way. When deep investigation in the past has led you to surprise, joy, or new inventions, you begin to crave deep investigation of anything valuable.

So, if you think mathematically, you don't just look at a news story, such as the amazing recent discovery of gravity waves, and say, "Ho-hum, that's interesting," while moving on to the next distraction. Rather, the gears start to turn in your brain. You begin to read the story in earnest, to place this discovery in the context of what you already know. You learn that gravity bends the geometry you learned in high school: while light still follows straight lines, gravity changes what *straight* means! You are fascinated and want to learn more. You begin to imagine a universe awash in gravitational ripples from events long ago, and the mathematical challenge of recognizing those events. You begin to deeply appreciate how these waves offer a new way to observe the universe. Going down this rabbit hole has given you a richer view of what is going on in the world.

The quest for deep knowledge in mathematics builds a virtue in us: *thinking for oneself.* If, as the writer Kenneth Burke once said, literature is "equipment for living," then mathematics is equipment for thinking.[6] You discern when an answer makes sense, or when something isn't quite right. You are not resigned to a life of conformity and blind trust in authority. You are better able to tell if someone is trying to fool you. And mathematical reasoning and sense making cultivate in us the virtue of *thinking rigorously:* the ability to handle ideas well and to craft clear arguments with those ideas. This virtue serves us well in every area of life. Of course, studying mathematics isn't the only way you learn to think for yourself or think rigorously, but it is one of the best ways to do so.

The quest for deep investigation in mathematics builds the virtue of *circumspection.* In mathematics, we are always quantifying our statements, domains, bounds, and assumptions so that our claims are in fact true. We are trained to know the limits of our arguments, and that helps us not to overgeneralize. In mathematical modeling—the process of describing a real-world problem mathematically—we state the assumptions of our model and its limitations. In statistics, we are careful to point out that correlation is not the same as causation. Can we not learn from these examples to be more circumspect in our language about people, and not make sweeping generalizations? I hope so. Of course, we all have unconscious biases that push us toward certain associations, but the math-trained person may be more equipped with examples that highlight the logical flaws in doing so.

The quest for truth in mathematics predisposes the heart to the virtue of *intellectual humility.* Isaac Newton said:

> I do not know what I may appear to the world; but to myself I seem to have been only like a boy playing on the seashore, and diverting myself in now and then finding a smoother pebble or a prettier shell than ordinary, while the great ocean of truth lay all undiscovered before me.[7]

He's saying: the more we know, the more we realize how much we do not yet know. That's a posture of humility. Math explorers are often paying attention to what they don't know, because unsolved problems are more interesting. Explorers are in the habit of making conjectures, many incorrect, so they learn how to accept blunders as a normal phase of exploration—in fact, being wrong is celebrated. Just look at the titles of math books like *Counterexamples in Topology* or *Counterexamples in Analysis.* Admitting when one's arguments are faulty is a skill I want to build in my students. This is the virtue of *admitting error.* In the past, when I've put a difficult question on an exam, some students would make stuff up, hoping for any partial credit. Now I explicitly reward students who can acknowledge where their reasoning has gaps—I get much more thoughtful answers that way.

Thus, when a math explorer seeks important truths, she demands deep knowledge, investigates deeply, embodies intellectual humility, and is willing to revise her beliefs in light of new information. She handles ideas rigorously, with honesty and integrity. She values circumspection and clarification of distinctions. She handles the truth accurately.

Unfortunately, these virtues may not be strong enough to overcome hardened worldviews or confirmation bias—the tendency we have to favor only information that supports our preexisting beliefs—especially when there is emotion or identity involved. So when truth is so obscure and falsehoods so established, how do we love the truth? Why should we bother to figure out what's true, or be willing to revise our beliefs?

When my dad had cancer, our family could not sit back and let someone else sort things out. We had to know what treatments had the best chance of saving his life. We sought expert opinions and combed through all relevant information. You see, truth matters to us when we realize that there's a lot at stake.

What's at stake for us if we don't seek deep knowledge and deep investigation of important truth claims?

At stake is our ability to not be manipulated or taken advantage of. At stake is our ability to make informed decisions. At stake is our ability to become innovators—not just consumers—of new technologies that could change the way we live. At stake is the ability to look at the ways technology is being used and to critique it. At stake is our ability to protect our loved ones from harmful falsehoods. At stake is our ability to dialogue with people who hold different beliefs.

As you grow as a math explorer and demand deep knowledge and deep investigation of every important truth, you'll learn more about the world and your place in it. You may see that some of the societal issues you thought were so clear cut are more complex than you realized.

But the quest for truth in mathematics builds in us a noble hope that we can, indeed, know the whole truth, even of messy, complex truths—and that some things are absolutely, fundamentally *true*—so that we have confidence in truth to combat those who wield lies brazenly. *Confidence in truth* is perhaps the most important virtue solidified by the quest for truth in mathematics. The more you explore the world of mathematics, the more confidence you have in loving the truth and knowing that it really is worth the effort to figure out.

VICKREY AUCTION

In game theory, there is mathematics that incentivizes people to tell the truth. Game theory is the mathematical modeling of decision making. It is useful for analyzing strategic thinking, and it has found diverse applications in economics and computer science. Here is one neat result from game theory.

Suppose you are hosting a silent auction to sell your car. The rules are (1) prospective buyers bid for your car by placing their bids in sealed envelopes, and (2) after collecting all the bids, you sell the car to the highest bidder for the price that he bid.

This may seem like a reasonable way to conduct a silent auction, but there is a risk involved for you, the seller. If your car is highly valuable but all the buyers think themselves the only one who recognizes this, they may make bids that are *less* than they think the car is worth. As a result, your bids will be lower, and you will suffer.

Is there a type of silent auction that will induce people to bid what they think an object is truly worth?

The answer is yes, and it is called a *Vickrey auction*. The rules specify that each prospective buyer makes a bid, and the car goes to the highest bidder, but *at the second-highest bid price!*

Can you figure out why this induces people to bid truthfully? In other words, why is telling the truth a better strategy than overbidding or underbidding?

Google Ads uses a generalization of the Vickrey auction to sell ads in sponsored searches.

September 5, 2018

About this draft you sent me . . . the last paragraph was particularly powerful, and it completed a lesson for me. My study of formal logic in mathematics is what gave me the trust in my reason to seek, recognize, and build a legal argument for my appeal (plus a piece of information I'd happened upon that caused me to investigate). And confidence in truth is often an impetus to act.

I can definitely relate to the paragraph about your experience with theorems. When I first encounter and then touch base with some mathematical concepts, I have the same experience, and it's part of what pushes me forward (the challenge too: I enjoy the "fight" or struggle).

Chris

8
struggle

Every time that a human being succeeds in making
an effort of attention with the sole idea of increasing his grasp
of truth, he acquires a greater aptitude for grasping it,
even if his effort produces no visible fruit.
Simone Weil

Practice is a means of inviting the perfection desired.
Martha Graham

My stomach sank as I read the two students' papers. Their math proofs looked unusually similar in notation and wording, yet I was quite sure neither student knew the other. On a hunch,

I searched online for the problem I had posed, and found a solution that was the likely source of both answers.

What should I do? Rather than confront the students and put them on the defensive, I decided to give them an opportunity to step forward. If they took responsibility for their actions, I could advocate for a less severe penalty before the college judiciary board. I emailed the entire class, explaining that I'd discovered instances of cheating using online resources but I was hoping that the responsible parties would step forward.

The next morning, I was startled to find ten confessions in my inbox, including the two I had identified. Though I'd heard of significant cheating incidents at other schools, I was surprised that this had happened in my class.

What had compelled so many students to cheat? Some confessed to feeling enormous pressure to achieve because they wanted the approval of peers or parents, or to get a good job or into graduate school. Said one, in tears: "I really *did* try the problem! But I was tired, and had so much work to do . . . so I looked for a solution online. I just wasn't sure if I kept at it that I would have solved it." It saddens me that now she'll never know if she would have solved that problem. She's cemented the very insecurity she was trying to destroy.

Is the Internet to blame? Yes and no. The underlying temptations have always been there. But the Internet has greatly amplified our ability to compare ourselves to others and lowered the barriers to indulging our immediate desires—even ones that aren't good for us. Twenty years ago, Facebook didn't exist, but now, with the explosion of sharing on social media, it is far easier to feel inadequate when we see only highly curated versions of everyone else's lives and accomplishments. The pressure from social comparison has never been greater. It's also easy now to find solutions to any math problem online. In one advanced course, I've noted a decline in the number of students who come to see me about the most difficult problems, compared to just a few years ago. It's too easy now to avoid the path to learning that comes through struggle.

But why struggle? What is the value of struggle? How is struggle a deep human desire?

One kind of struggle is the struggle through suffering. In life, most of us don't go looking for this kind of struggle. We don't relish suffering. Yet suffering is a reality of the human experience. Many of our most vivid experiences come through our own suffering, or walking with a loved one in their struggle. Those who have suffered know that suffering produces endurance, and endurance produces character, and character builds hope. The struggle, whether against illness or injustice, builds virtues that cannot be taken away from us by disease or by force. Such virtues are important for a well-lived life. But this kind of struggle, as valuable and as commonplace as it is, is not a deep human desire.

Another kind of struggle is the struggle to achieve. But with what goal? After all, if the goal is achieving a good grade, one can just avoid that struggle by cheating. In the book *After Virtue*, the philosopher Alasdair MacIntyre distinguishes between *external goods* and *internal goods* in any social practice. As he defines it, a *practice* is, roughly speaking, a socially established cooperative human activity with standards of excellence appropriate to that activity. It could include things like sport, farming, architecture, mathematics, or chess. External goods are goods that come about by engaging in the practice, but only as accidents of social circumstance and not inherently because of the activity; for instance, wealth and social status are external goods. And since they aren't inherent to the activity, external goods can always come about another way. MacIntyre gives an example of a child incentivized with candy to play and win at chess:

> Thus motivated the child plays and plays to win. Notice however that, so long as it is the candy alone which provides the child a good reason for playing chess, the child has no reason not to cheat and every reason to cheat, provided he or she can do so successfully.[1]

By contrast, internal goods are goods that come about by engaging in a practice and are bound up intrinsically with that practice; such goods cannot be had apart from engaging in that or a similar practice. MacIntyre continues:

> But, so we may hope, there will come a time when the child will find in those goods specific to chess, in the achievement of a certain highly particular kind of analytic skill, strategic imagination and competitive intensity, a new set of reasons, reasons now not just for winning on a particular occasion, but for trying to excel in whatever way the game of chess demands. Now if the child cheats, he or she will be defeating not me, but himself or herself.[2]

Thus "strategic imagination" is an internal good. You develop it by playing chess or a similar game, and it is tightly bound up with, and a result of, the activity.

As MacIntyre points out, external goods (candy, wealth, fame, etc.) belong to the individual, and it's often the case that the more that goes to one, the less is available for others. By contrast, internal goods (excellence at a skill, the joy of an activity) can accrue to any number of individuals without diminishing the amount still available to others, and such goods enrich the whole community of people who participate in the practice. Your acquired skill at mathematics is good for society as a whole. Your mathematical discoveries and statistical insights may have applications that benefit everyone. Beyond that, every mathematical theorem, definition, and application stands as a testament to human ingenuity and is a credit to us all.

Thus, the struggle that I want to amplify as a basic human desire is the struggle to achieve internal goods—perhaps more succinctly described as *the struggle to grow.* Everyone has an assortment of social practices they engage in, such as sport and work and friendship. In every domain we have a basic human desire to grow, to fulfill the potential we have in those areas. When I exercise, I am striving toward the internal goods of fitness and staying healthy (and at my age, I am less interested in the external goods that come from working out, such as big muscles or social status). Everyone I know has a desire for meaningful work, and this likewise comes out of the basic human desire to grow in our vocation, for professional development and personal satisfaction. And each of us has a desire for deep and rich and meaningful friendships that we can grow with and into.

The struggle to grow—to achieve internal goods—is a mark of human flourishing. A society that does not structure its social practices to promote internal goods has no anchors. For instance, education is a social practice that builds, among many internal goods, the ability to think critically. An external good that education offers is the appearance of authority. But a society that does not encourage critical thinking is easily swayed by propaganda and misinformation from actors who have gained the appearance of authority by other means. Moreover, a declining society—with deep inequity, lack of opportunity, or corrupt leadership—incentivizes people to acquire external goods dishonestly, because they see their leaders unfairly distributing them. However, internal goods are intrinsic to social practices, so they cannot be obtained fraudulently. Pursuing them requires and cultivates virtues. When people appreciate the inherent value of internal goods, they see the struggle to grow as a means of flourishing—even within an unjust system—and as a means of fighting against it.

The struggle to grow is an attraction of the mathematical experience. Math explorers relish interesting puzzles and hard problems. We know what it is like to struggle with a problem for a long time and possibly get nowhere with it. We learn to *enjoy* the struggle. In my own math research, I've spent *years* thinking about some problems. It's possible that I will never solve those problems, just like I may never solve some of my life's problems. That only makes the joy so much sweeter when an insight comes and I finally solve one of them.

In the math education community, the term *productive struggle* describes the state of actively wrestling with a problem, persistently trying out various strategies, being willing to take risks, being unafraid of mistakes, and progressing incrementally in understanding the underlying ideas. This wrestling produces a certain kind of *endurance,* which enables us to be comfortable with the struggle. This endurance produces an *unflappable character* that benefits us in addressing life problems—calming us with the knowledge that it's okay if we don't solve a problem right away. We appreciate that not solving a problem can be just as important as solving it—that, as Simone Weil suggested, the effort to grasp truth is itself worthwhile, for increasing our aptitude, even if it produces no visible fruit. In the struggle, we acquire *competence to solve new problems,* and fortify an expectant hope that we will one day solve them. And when we struggle, and at last succeed, we build *self-confidence.* Over time, through incremental and hard-won victories, this leads to *mastery.*

These are all virtues that are cultivated through the proper practice of mathematics, the kind that emphasizes internal goods over external goods—the kind that taps into every person's deep desire to grow through struggle. So what can math explorers do to encourage this kind of struggle, while discouraging the temptation to circumvent the struggle and succumb to the allure of external goods?

In my reflection on the cheating incident, I faced a related question: in what ways was I responsible for what had happened? I began to feel disappointed, not so much in my students but in myself. Was I inadvertently causing my students to feel that grades were paramount? What could I do differently?

According to research on academic dishonesty, rates of cheating have risen over the past several years, and technology has contributed.[3] Moreover, one of the strongest predictors of cheating is when parents or teachers place an undue emphasis on grades. Students are less likely to cheat in situations where teachers emphasize mastery—learning for its own inherent value—over external outcomes such as grades. Again, we see the distinction between mastery, an internal good, and the achievement of satisfactory grades, an external good.

Whether we realize it or not, there are subtle ways that we signal the importance of achievement, even if we value mastery. For instance, whom do we praise in the home or in the classroom? On whom do we shower attention? If we show more fondness for A students than for C students, we are implicitly valuing performance over mastery. Even if we are not sending such messages, kids receive signals from society that grades matter, and they are prone to blowing those signals out of proportion. We must take steps to actively counter the idea that grades are supremely important.

Because of the cheating incident, I now ask students to read a short article by Carol Dweck about her research showing that people who believe intelligence is fixed ("a fixed mindset") are more fearful of challenges and more easily discouraged by failure than people who believe that intelligence is malleable and can grow ("a growth mindset").[4] Students with a fixed mindset equate talent with doing things easily. Consequently, they see struggling with problems as evidence that they do not have ability. By contrast, those with a growth mindset see that setbacks are an opportunity to learn and can be overcome with perseverance through struggle.[5]

These accomplished mathematical thinkers, all three Fields Medalists, emphasize the value of struggle in mathematics:

> I am a slow thinker, and have to spend a lot of time before I can clean up my ideas and make progress.—Maryam Mirzakhani[6]

> It is terribly important to have the experience of finding things difficult but then managing to get past that stage. If the habit of thinking for yourself and solving problems even though they are difficult could be instilled at a very early stage, it would make a huge difference.—Timothy Gowers[7]

> In spite of my success, I was always deeply uncertain about my own intellectual capacity; I thought I was unintelligent. And it is true that I was, and still am, rather slow. I need time to seize things because I always need to understand them fully.—Laurent Schwartz[8]

I remind my students continually that struggling is a good thing, that it's where learning happens: it's what professors are always doing in our research, and the struggle is the most interesting place to be. I remind them that they are building virtues that will serve them well in any of life's challenges, because they will know how to persist, through difficulty, to reap the rewards on the other side. I remind my students that grades are a measure of progress, not a measure of promise. They do not determine your dignity as an individual.

In reflecting on these things, I began adapting my assessments to reflect how I value the struggle. I now give partial credit to students who can show me they've thought through a strategy, even if they couldn't solve the problem. I ask reflective questions, like this one, that show I value the process of doing mathematics, not just the outcome:

> Reflect on your overall experience in this class by describing an interesting idea that you learned, why it was interesting, and what it tells you about doing or creating mathematics.

The responses are often a joy to read. Like this:

> I've learned a lot of interesting things in this class, but I think my experience is best reflected by the article we were instructed to read for Homework 0. This article talked about how [some] think of intelligence [as fixed,] so that when things get difficult, one feels helpless. . . . Until recently, I was often frustrated by math and unhappy with myself because I held a view similar to this. In the article, the alternative is to understand that learning takes effort and persistence. While this is not a great revelation to me, I feel that I have learned to better accept it this past semester. This has been an important lesson to me because I now have more confidence to continue with math. I have learned that doing and creating mathematics can depend on insight and inspiration, but [are] also largely dependent on putting in the effort to become good at things.

Here's another question I asked recently:

> One of the luxuries of the Internet era is that you can look up the answer to almost any problem you want—as long as it's been solved. Yet when you are learning a subject it can be counterproductive. In this class, I have emphasized the importance of struggling in mathematics: that it's normal and part of the process of learning, and that when you are stuck, you should just "try something." Describe an instance, so far in the course, where struggling and trying something was valuable to you.

The following response came from a former Marine who had returned to finish college:

> I know that learning to do something with your hands is probably the equivalent to struggling to do something academically. For example, when I was ten I taught myself to juggle and got really good at it. I stopped juggling once I realized it wasn't cool but 15 years later was able to shock my wife with this hidden skill and realized I could still do it. Similarly, a lot of the skills I learned in the Marines required me to perform some complicated action with my hands or body. I'm pretty sure I will always remember how to disassemble and assemble any crew-served machine gun. Similarly, shooting is something I spent years doing and I'll probably always be good at it until my body stops working.
>
> I look at math professors effortlessly solve problems on the board even if they didn't initially remember how to do the problem. I've seen you do this at least a couple times when a student asked a good question. I think that for you math is a lot like fixing a car or putting something

> together without instructions. You have struggled so much that all previous experiences are burned into your brain. You simply cannot forget things that easily anymore, so because of this you are just better at math because you put in the time and hard work.
>
> I'm hoping that I can get to that level with some subject someday.

Seeing his understanding of the value of struggle, I am confident he will.

PENTOMINO SUDOKU

Here's another unusual sudoku puzzle courtesy of Philip Riley and Laura Taalman of Brainfreeze Puzzles, from their book *Double Trouble Sudoku.*[a] It's definitely harder, and you'll struggle a bit. All of the usual sudoku solving techniques get broken, due to the fact that each row and column contains the same set of numbers *twice.*

The board is divided into regions with five squares each (pentominoes). The goal is to fill each square with a single number, chosen from 1 to 5, according to the following rules: (1) each pentomino must contain the numbers 1 to 5 exactly once, and (2) each row and column must contain the numbers 1 to 5 exactly *twice.*

The shading in this puzzle has no purpose other than to group pentominoes of the same shape.

<table>
<tr><td></td><td></td><td>4</td><td></td><td>2</td><td></td><td></td><td>2</td><td></td><td></td></tr>
<tr><td>1</td><td></td><td></td><td>3</td><td></td><td></td><td>5</td><td></td><td></td><td>1</td></tr>
<tr><td></td><td>2</td><td></td><td>2</td><td></td><td></td><td></td><td></td><td>4</td><td></td></tr>
<tr><td>3</td><td></td><td>4</td><td></td><td>2</td><td></td><td></td><td></td><td></td><td></td></tr>
<tr><td></td><td>2</td><td></td><td>3</td><td></td><td></td><td></td><td>5</td><td></td><td>4</td></tr>
<tr><td>2</td><td></td><td>2</td><td></td><td></td><td></td><td>4</td><td></td><td>3</td><td></td></tr>
<tr><td></td><td></td><td></td><td></td><td></td><td>4</td><td></td><td>4</td><td></td><td>2</td></tr>
<tr><td></td><td>1</td><td></td><td></td><td></td><td></td><td>2</td><td></td><td>2</td><td></td></tr>
<tr><td>5</td><td></td><td></td><td>4</td><td></td><td></td><td>3</td><td></td><td></td><td>1</td></tr>
<tr><td></td><td></td><td>2</td><td></td><td></td><td>4</td><td></td><td>1</td><td></td><td></td></tr>
</table>

a. Philip Riley and Laura Taalman, Brainfreeze Puzzles, *Double Trouble Sudoku* (New York: Puzzlewright, 2014), 189.

August 9, 2018

I'd say I'm drawn to mathematics because of its strength, its structure, and its truth. To me, I've never seen an argument stronger than a mathematical argument (vs. philosophical, legal, etc.). Its structure is amazing, the least being how you can come from a variety of valid methods and still lead to the absolute same conclusion. And its truth [is amazing], being as how it describes our physical world (physical world being everything we know to exist), and its pervasive application to seemingly everything else. An example of the strength of mathematics is in our universe: it seems "mathematical physics" discovers most things about our universe before "real" physics does.

Chris

9
power

Power does not corrupt men; fools, however, if they get into a position of power, corrupt power.
George Bernard Shaw

The moving power of mathematical invention is not reasoning, but imagination.
Augustus de Morgan

Consider a standard deck of fifty-two playing cards. This is an everyday item that many people use for games or magic tricks, but rarely give a second thought. But mathematics equips you to see the deck in fresh, powerful ways.

The cards in a deck are arranged in some (not necessarily special) order, which we'll call a *configuration* of the deck. One of the first questions you might ask about a deck is "How many possible configurations are there of a deck of fifty-two cards?" Some of you know the answer, but since the heart of mathematics isn't calculation, let me ask you a different question. For this, I encourage you to go with your gut reaction, and don't try to calculate. Which of these quantities is the largest?

A. the number of stars in the universe
B. the number of seconds since the big bang—the beginning of time
C. the number of possible configurations of a deck of fifty-two cards

Now *that* is a more interesting question, which I leave you to ponder for a moment.

As we'll soon see, understanding a deck of cards can reveal many ways that math is powerful. Such powers become ours when we take up the practices of mathematics to unlock and expand our innate capacity for reason.

Power is a universal human desire. Yet *power* often sounds like a bad word, and powerful people are not universally admired. To understand why, we need to untangle two ways we speak about power.

The first concerns the power of *things*—like electric power or a powerful thunderstorm. The word *power* has its origins in the Old French word *poeir* and the earlier Latin word *potere,* from which we also get the words *potent* and *potential.* So a powerful thing has the capacity to do something. Math explorers often speak of mathematics as powerful in this way.

The second concerns the power of *people* to direct or influence other people or events. People can do many good things with power, but unfortunately, it's not always the case that they do. When power is abused, the dynamics of that power can negatively affect the way that mathematics is taught and learned. The sociologist Max Weber defined *power* as the ability of a person to impose his or her will upon others, even against resistance.[1] Weber was referring to the power of coercion, the ability to force someone to do something. This kind of power does not lead to human flourishing, for either the perpetrator or the victim.

I prefer a different way of thinking about power, which captures the best kind of power, both of things and of people. The writer Andy Crouch offers this definition:

> Power is the ability to make something of the world . . . the ability to participate in that stuff-making, sense-making process that is the most distinctive thing that human beings do.[2]

Two phrases here, both purposefully nebulous, need unpacking: *stuff making* and *sense making.*

Stuff making refers to not just the potency of humans, but the potency of the world. Electric power as a form of energy is used to make stuff. People make stuff and change their surroundings. Mathematicians have a nice word for this: *transformation.* Just as mathematical functions transform the things they operate on, creatures of the world are transforming their environments. The universe itself is in a constant state of transformation.

Sense making describes the ability to understand the world and, in its fullest expression, to make meaning of the world. It's also creative—it requires imagination. People make sense of the world. Objects do not make sense of the world, but some things, like mathematics, help people to make sense of the world.

Math explorers do both of these things. We make stuff—we

make definitions, we create structures, we prove theorems, we develop models—but we also make sense—the models and symbols we create are invested with meaning. By contrast, a computer might be able to participate in stuff making—performing tasks to compute answers—but it does not (as of yet, anyway) participate in sense making.

Crouch argues that stuff-making, sense-making power—*creative power*—is the deepest and truest form of power. It is a sign of human flourishing, and it does not always look like the power we are used to. Babies have power: the ability to participate and grow in stuff making and sense making. Mother Teresa had power: through her care for the poor, to make mutual meaning with the people she cared for.

However, creative power can be distorted, because humans have the potential to use their creativity for harm. When distorted, creative power becomes *coercive power.* Coercive power undermines the creative power of others to participate in stuff making or sense making.

Both creative and coercive power exist in mathematical spaces: environments where mathematics is done. Let's first examine the creative power of mathematics, and witness its sense-making capabilities.

What does it mean to say that *math is powerful?*

Let's return to card shuffling, and in this setting I'll illustrate all the ways that math showcases its power. As I do, let me just say that I'm trying to give you a sense of what's powerful about math, so as we walk through this don't worry if you don't understand everything. It's certainly okay—in fact it's normal, even for mathematicians—not to follow every detail the first time around. We are just flying over the landscape, and it's okay to simply enjoy the sights from a 50,000-foot view.

We began looking at our deck through mathematical eyes by asking a comparative question: which is largest—the number of stars in the universe, the number of seconds since the big bang, or the number of possible configurations of a deck of fifty-two cards?

This is a far more interesting question than just asking "How many ways are there to order a deck?," because the large numbers we are comparing are now attached to meanings. That points to a very basic power of mathematics: the power of *interpretation.* A math explorer doesn't stop when she does a calculation, because math is about comprehension, not calculation. A math explorer will look at the result of a calculation and try to interpret what it means, to see if it makes sense and how it fits in with other things she knows.

Astronomers estimate the number of stars in the universe to be about 10^{23}. That's twenty-three copies of 10 multiplied together, a number consisting of a 1 followed by twenty-three zeroes. Evidence from astronomy also suggests that the universe is about 13.8 billion years old, or less than 10^{18} seconds. The number of ways to order a deck of fifty-two cards can be computed by looking at the number of ways to choose the first card (52) times the number of ways to choose the second card (51, after the first card is fixed) times the number of ways to choose the third card (50, after the first two cards are fixed) times . . . etc. This product, when you multiply all the whole numbers from 52 down to 1, is called "52 factorial" and is written "52!," where the exclamation point stands for "factorial." (So, for example, 5! is $5 \times 4 \times 3 \times 2 \times 1 = 120$.) Using a factorial symbol always makes us look excited, and for good reason: the number 52! turns out to be about 10^{68}, a mind-bogglingly large number of different ways to order a deck of cards! This is way more than either stars in the universe or seconds since the big bang. And, if you take a moment to exercise your power of interpretation, you'll realize:

> If a deck of cards were shuffled once per second since the beginning of time, it would still be nowhere close to realizing all the possible configurations of cards.

In fact, 10^{68} is so much larger than 10^{18} that each time you shuffle a deck, it is exceedingly likely that the resulting configuration has never happened before in any deck that's ever been shuffled. In other words:

> Each time you shuffle a deck of cards, you are making history![3]

A second question that's natural to ask is "How many shuffles does it take to mix a deck of cards well?" Let's restrict our attention to the *riffle shuffle*, which is the shuffle that most people do, where you cut the deck and shuffle the two halves together in a more or less interleaved way.

You may have heard that it takes seven riffle shuffles to mix a deck of fifty-two cards. That is a theorem published by the mathematicians Dave Bayer and Persi Diaconis in 1992,[4] and I'll give the general thread of their argument here.

To start, the question of "how many shuffles" has some ambiguity. What does "mix well" mean? This is another power of mathematics: the power of *definition*. Math explorers try to make precise those words that aren't yet well defined. Before we talk about mixing, we must first describe the "state of our knowledge" of the deck: the probability of the deck's being in this or that configuration. A "probability distribution" tells us how likely any configuration of cards is. As we've seen, there are

52!, or about 10^{68}, configurations, so a probability distribution must specify the probability of every single configuration—if you forced me to write this out in a list, it would be very long, with 52! rows, and each row would list a configuration together with the corresponding probability.

At the start, before any shuffles, the deck can be in only one configuration, so the probability of that configuration is 1 and the probability of all others is 0. Then shuffling injects some randomness. After shuffling, the state of the deck has become more uncertain. Some configurations are more likely than others, and a probability distribution tells us the probability of each. The ultimate example of a well-mixed deck is one for which every configuration has the same probability, so that we have no special knowledge of the state of the cards. We can exercise our power of definition to give this a name.

Let's call a deck "random" if *every* configuration is equally likely. A random deck is really a probability distribution which says that each of the 52! configurations of the deck has a 1/52! chance of happening. So to measure how well-mixed a deck is, it makes sense to quantify, if we can, its "distance" from a random deck. This is another potent aspect of mathematics: the power of *quantification.*

Now, what do we mean by "distance"? Distance in what space? Here is where math shows several more of its powers: the powers of *abstraction, visualization,* and *imagination.* We are going to use our imaginations to visualize an abstract space where every point is a probability distribution, so that each point represents a state of knowledge of the deck (see facing page). The random deck (no knowledge) will be one point in the space, and other points in the space will be other probability distributions (other states of knowledge). We will want to know how far the state of our knowledge is from the random deck after each shuffle.

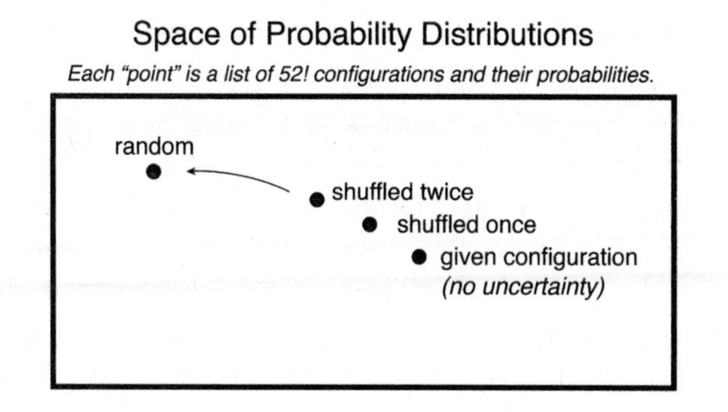

Then we are going to create a function to measure the distance between points in the space of probability distributions, just like you might do for points in a real space. Here we get to exercise our power of *creation*. Yes, this requires some creativity, because there are many options you could pick. For instance, if I wanted to measure a "distance" between two people in the world, there would be lots of choices—here are a few:

1. physical distance, in miles
2. friendship distance (or degrees of separation), the shortest chain of friendships between two people
3. distance in flight time, the shortest time via plane flights
4. distance in car/boat time, the shortest time via surface transportation
5. genealogical distance, how many generations back to a common ancestor

You can probably think of others. To decide among them, we would exercise our power of *strategization*. Math explorers learn how to make strategic choices in problem solving. A popular myth of doing math is that either you see the answer or you don't. In reality, you demonstrate mathematical power by assembling possible strategies and trying them to see if any will work.

For card shuffling, we are in a space of probability distributions. The notion of distance that Bayer and Diaconis chose for this space is something called "total variation distance." What that is won't concern us here, since we are speaking only in big-picture terms, to give you a feeling for this result. They chose this distance because it has good properties for measuring how far a shuffled deck is from a random deck after each shuffle.

So we need to analyze shuffling and what it does to probability distributions. For this we draw on another math power: the power of *modeling*. We set up a mathematical model for shuffling and hope that it accurately represents how people shuffle. The shuffling model that Bayer and Diaconis chose is called the Gilbert-Shannon-Reeds shuffle (or GSR shuffle), which is a pretty good approximation of how people actually riffle shuffle. The model assumes that the deck will be cut "binomially," which means that the deck will be cut into a packet of size k and a packet of size $(52-k)$ with the same probability as k heads showing up after flipping fifty-two fair coins. Then cards are dropped, in succession, from one packet or the other with probability proportional to the number of undropped cards currently on either side. Note how the mathematical description of this shuffle contains some randomness, because people don't shuffle the same way every time. For instance, the deck won't be cut exactly in half all the time, but the packets are likely to be nearly the same size, just as flipping fifty-two coins is likely to yield approximately the same number of heads as tails.

This GSR shuffle description may seem unwieldy. But as Bayer and Diaconis showed, there are at least four other equivalent descriptions of the GSR shuffle. One is a geometric description that moves cards around in a way similar to kneading dough. Another is an entropy description, which says that all possible pairs of cuts and interleavings are equally likely (thus, more lopsided cuts—which have fewer interleavings—are less likely). The fact that there are multiple ways to describe the GSR shuffle highlights the power of *multiple representations* in mathematics—if there are multiple ways to understand an idea, you have the power to choose the one that makes the problem easiest to solve with the tools that you have.

These representations are all best understood by working with a generalization of the riffle shuffle called the n-shuffle. This is like a riffle shuffle, but instead of cutting a deck into two portions, you cut a deck into n portions before interleaving them. The power of *generalization* in mathematics is that often, solving a general problem can yield more insight than solving a specific problem. An m-shuffle followed by an n-shuffle turns out to be equivalent to doing one mn-shuffle. So doing a regular riffle shuffle twice is doing a 2-shuffle followed by a 2-shuffle, which is equivalent to doing a 4-shuffle. Follow that with one more riffle shuffle and you've done the equivalent of one 8-shuffle. Similarly, riffle shuffling a deck k times is equivalent to doing one 2^k-shuffle.

This useful idea helps us see an algebraic structure to shuffling—highlighting the power of *structure identification* in mathematics. Mathematics helps reveal structure that we didn't see before, and that structure can charm us as well as suggest how to solve our problem. Another structure in the deck is its so-called *rising sequences*. Within a deck splayed out from left (bottom card) to right (top card), a rising sequence is a maximal subset of cards consisting of successive face values displayed in order. For instance, this deck of ten cards has three rising sequences:

That's because {A, 2, 3} is a rising sequence, {4, 5} is another, and {6, 7, 8, 9, 10} is another. Within a rising sequence, the face values must be successive (the way they appeared in the original, ordered deck) and the successive sequence must extend as far as possible in the current deck. So {6, 7, 8} is not a rising sequence here, because it can be extended to {6, 7, 8, 9, 10}.

A completely ordered deck has just one rising sequence:

Let's see what happens if we give this deck a riffle shuffle. First we cut the deck binomially, which means that it will split into packets of size six and size four with the same probability that six heads show up in ten fair coin tosses:

Then we riffle the packets together by dropping cards from each packet with probability proportional to the packet's size. Thus the ace has a 6 in 10 chance of dropping first, and the 7-card has a 4 in 10 chance of dropping first. Suppose the ace drops first. Then the next card that drops will be either the 2-card, with probability 5/9, or the 7-card, with probability 4/9, since the packets have sizes five and four now. Suppose the 7-card drops next. Continuing, we might obtain this interleaving:

You'll see that this shuffled deck has two rising sequences: {A, 2, 3, 4, 5, 6} and {7, 8, 9, 10}. If you shuffle once more, each of these two rising sequences may get split when the deck is cut, unless it is a very lopsided cut. Then, when you interleave the cards and finish the shuffle, you'll have at most four rising sequences (fewer if the cut or interleaving was lopsided). For similar reasons, if you shuffle a third time, you will have at most eight rising sequences. (In fact, we noted earlier that shuffling three times is equivalent to an 8-shuffle, and the rising sequences will arise from the eight packets in the 8-shuffle.)

Knowing about rising sequences helps us see that not all configurations are possible after three shuffles, because there are some configurations of these ten cards that have *more* than eight rising sequences. In fact, the reversed deck has ten rising sequences, because you need to pass through the deck ten times from left to right to hit all the cards in order (since each pass hits only one additional card).

You can use a similar argument (try it!) to argue that for a deck of fifty-two cards, some configurations cannot be reached after five shuffles. So, with only the power of our minds, we can establish that five shuffles are *not* enough to mix the deck, because that number of shuffles can't even reach every configuration, much less make them all nearly equally likely.

It should surprise us, then, that only two shuffles later, every configuration is nearly equally likely! Bayer and Diaconis discovered that the probability that an n-shuffle will result in a particular configuration depends only on the number of rising sequences, the number of cards, and the number n. From this, they could calculate the total variation distance between a deck shuffled with a 2^k-shuffle (equivalent to riffle shuffling k times) and the random deck. Their analysis shows that for fifty-two cards, you need at least seven shuffles to be close to a random deck, and while more than seven shuffles takes you closer to random, it does not do so in an appreciable way. So, in that very quantifiable sense, seven shuffles is the right number of

GSR shuffles to mix a deck so that every configuration is nearly equally likely.

There are a number of things to reflect on in this result. One is that it should be very surprising that we can say anything so precise about a deck with more configurations than stars in the universe! Another surprise is that even with so many configurations, it takes only seven shuffles to make them all nearly equally likely!

In this one extended example, we've now seen many of the powers of mathematics—interpretation, definition, quantification, abstraction, visualization, imagination, creation, strategization, modeling, multiple representations, generalization, and structure identification. Anyone who learns mathematics will grow skilled in these powers. These are virtues that enable the creative power of stuff making and sense making.

But just like any good thing, mathematics can be used in harmful ways, and its powers can be distorted to become coercive, because people are fallible.

Coercive power disrupts other people's capacity to exercise creative power. When we deny others the opportunity to get a good education, we deprive them of tools for stuff making. When we avoid working with "difficult" students, we have harmed their ability to flourish. Coercive power limits other people's ability to create meaning for themselves or their work. When we exclude people because of their race, gender, religion, sexual orientation, class, or disability, we are preventing them from being meaningful participants in society.

It was not long ago that women had to overcome active opposition from universities in order to pursue mathematics. Sofia Kovalevskaya (1850–91) is a Russian mathematician known for important theorems in partial differential equations regarding the propagation of heat and the motion of rotating bodies. However, as a student in Saint Petersburg and Heidelberg, she was able to attend university classes only unofficially. When she moved to Berlin to study with the famed mathematician Karl Weierstrass, the university there refused to let her attend his classes at all, despite his petitions on her behalf. So he tutored her privately. When she had done enough research for her doctoral dissertation—which includes one of the results for which she is known today—they had to search for a university that would give her a degree. The University of Göttingen awarded her a PhD, in absentia, in 1874. She was the first woman in the world to receive a doctorate in mathematics. Even though one of her PhD results was published in the most prestigious German math journal at the time, she was unable to get a job, in either Germany or Russia. She left mathematics and wrote fiction and theater reviews instead.[5] We might not have benefited from her subsequent outstanding contributions in mathematics if she hadn't returned to it six years later, and tried again.

Kovalevskaya's story shows that coercive power can hide inside social norms—because "that's the way it's always been done." It's easy to dismiss her example by saying that women don't face such obstacles anymore, but we would do well to consider what social norms we allow today, implicit and explicit, that are impediments to anyone. We should think about our own capacity to change such norms with the power we have.

Think about all the people who *haven't* prevailed against the people and institutions that have limited their creative power. Erica Walker's book *Beyond Banneker* recounts stories of African American mathematicians in the past century—many whose talent might have been overlooked or suppressed if not for serendipitous opportunities or the presence of advocates helping them to persist in the field.[6] Even today, barriers remain for women, people of color, and other disadvantaged groups to fully exercise their creative powers in mathematics.

Coercive power isn't always imposed by people—it can sometimes exist in structures that wield power in unseen ways. Poorly designed furniture can prevent those in wheelchairs from participating in the same ways that others can. Unnecessarily restrictive prerequisites might prevent low-income students from taking advanced courses—even though they might be ready for them—just because they didn't have the same depth of mathematical opportunities in high school as others. The algorithms that society increasingly relies on to "score" people for jobs, creditworthiness, or promotions may unintentionally reinforce bias if they are not designed thoughtfully and monitored accountably.[7] So we must reflect, not just on how we use power, but on how we cede our power to other things.

Creative power distinguishes itself in many ways from coercive power. Creative power *amplifies power in the subject and the object of power.* Think about what happens when you teach someone a new mathematical skill. Now more than one person has that skill. You've just magnified someone else's capacity for stuff making and sense making. You are also growing and becoming more skilled in your own power. Using mathematics to serve others also has an amplification effect. Solving any of the world's big problems (curing cancer, ending hunger, stopping human trafficking, etc.) will no doubt involve math (through mathematical thinking, mathematical modeling, or math-assisted innovations) and will rescue the creative power of thousands. Consider, too, the power of a kind word of encouragement. It takes nothing away from you, yet it lifts up and encourages another. Creative power is *humble,* and it puts others first. It seeks to unleash creativity in others. Coercive power would never do that. When students aren't performing well, instead of asking, "What's wrong with them?," a humble math teacher will ask, "What could I be doing differently?" Creative power is *sacrificial.* Parents expand the creative power of their children by spending time working with them. When you pursue this kind of creative power, you acquire the attendant virtues as well: having a *humble, sacrificial, encouraging character* with a *heart of service* and a *resolve to unleash creativity in others.*

Along these lines, the education activist Parker Palmer offers this wisdom:

> Teachers possess the power to create conditions that can help students learn a great deal—or keep them from learning much at all. Teaching is the intentional act of creating those conditions, and good teaching requires that we understand the inner sources of both the intent and the act.[8]

To use, teach, or learn math effectively requires careful thinking about power dynamics: how people interact, who has authority, who has freedom and who is constrained, who is encouraged and who is shut down, who is included and who is excluded, implicitly or explicitly. These are all questions of power. If there is any simple criterion to guide your actions with respect to power, it is this virtue: creative power *elevates human dignity.* What does it mean to "empower" someone? It means to affirm their dignity as a creative human being.

In a broken world, we must also realize that power is often unearned, and it doesn't just come to those who would use it well. So, you and I have a responsibility, as we grow in power, to use it for good—to grow in creative power and not coercive power. Creative power isn't solely instrumental. You don't grow in creative power only to accomplish things. You do it to be a better human being, to grow in virtue, to elevate human dignity—your own and that of those around you—in all mathematical spaces.

POWER INDICES

Power in political systems affects our daily lives, so it shouldn't be surprising that mathematicians and political scientists have developed models for quantifying power. One such model is the Shapley-Shubik power index.

Suppose you have a decision-making body of 100 people, consisting of group *A* (50 people), group *B* (49 people), and group *C* (just 1 person). To pass a bill, 51 people are needed, but the groups vote as a bloc. If you think about it, *C* can wield quite a bit of influence over the outcome, even though it's just 1 person. Such a situation happened in 2017 in the US Senate when Senator John McCain saved President Obama's health care law after 50 senators announced that they would oppose the repeal and 49 announced that they would vote for the repeal.

One way to quantify that influence is to imagine the voting groups entering a room in some order and forming a growing coalition; when the coalition is just large enough to pass a bill, the voting group that just entered is called the *pivotal group.* The Shapley-Shubik index of a voting group is the fraction of orderings for which that group is pivotal.

In our example, there are six orderings of the three groups: *A**B**C, AC**B**, BA**C**, B**C**A, C**A**B, C**B**A* (the pivotal groups are boldfaced). For example, group *A* is pivotal in four orderings, including *BAC* (because *B* doesn't have 51 votes by itself) and *BC**A*** (because *B* and *C* together do not have 51 votes). Group *B* is pivotal only in *A**B**C*, and *C* is pivotal only in *AC**B***. So the Shapley-Shubik index for group *A* is 4/6, for group *B* is 1/6, and for group *C* is 1/6. According to this measure of power, the 1 person in group *C* wields the same power as the 49 people combined in group *B*.

What power indices do you get if group *A* has 48 people, group *B* has 49, and group *C* has 3? Try it out. This would be another way to analyze what happened in 2017 if you want to think of Senators Susan Collins, Lisa Murkowski, and McCain as a coalition that voted outside their party's voting bloc to stop the repeal of the health care law.

In their book *Mathematics and Politics,* the mathematicians Alan Taylor and Allison Pacelli analyze the power of the US President (in the federal system that includes the House and the Senate) and find it to be about 16 percent. You'll also find there a discussion of political systems of other countries, and other notions of power.[a]

a. Alan D. Taylor and Allison M. Pacelli, *Mathematics and Politics: Strategy, Voting, Power, and Proof* (New York: Springer, 2009). The example with groups of sizes 49, 50, and 1 appears in Steven Brams, *Game Theory and Politics* (New York: Free Press, 1975), 158–64.

June 3, 2018

I know by now that you had tried to e-mail me a few times by now probably, but I've been in the SHU (Segregated Housing Unit) (the hole)[a] since April 16: I've been writing administrative grievances against the administration since I've been [at this prison]. I don't know if that's the reason why or if it's something else, but the administration has fabricated an incident against me and put me here.

Chris

a. This is a colloquial expression for solitary confinement. For several months before this, Chris and I had been using the prison's limited e-mail system to communicate, and when I didn't hear from him for many weeks, I began to get worried. Once he got in touch via this letter, we resumed corresponding by regular mail. Chris spent five months in solitary confinement, mostly cut off from other inmates. In August 2018 he went on a twenty-six-day hunger strike to protest his treatment. The prison where this occurred was not Pine Knot or the facility where he is currently detained.

10
justice

Justice. To be ever ready to admit that another person is something quite different from what we read when he is there (or when we think about him). Or rather, to read in him that he is certainly something different, perhaps something completely different from what we read in him. Every being cries out silently to be read differently.

Simone Weil

My favorite Chinese restaurant serves authentic cuisine just like my parents used to make. When you order an entrée, they bring you a little appetizer and a dessert as well! It's a bargain, so I don't complain that the appetizer (crunchy noodles) and dessert (Jell-O) are not themselves authentic.

But one day I went there with a Chinese-speaking friend. When the appetizer came, it was not crunchies but delectable pickled cucumbers. Yet my friend had made no special request. And when the dessert came it was red bean soup—my favorite when I was a kid! Why had I not gotten this before?

I began to see a pattern: when I came with non-Asian friends, I would get the crunchies and Jell-O. But when I came with Asian friends, I got the good stuff without even asking.

And then I noticed that my Chinese friends were also offered a completely different menu—a secret menu—with more-authentic dishes. I looked around the restaurant and beheld a bizarre sight: people side by side in the same space having very different experiences—non-Asians ordering from a standard menu and getting Jell-O, but Asians ordering from the secret menu and enjoying red bean soup.

"You won't like the stuff on that menu," I was told. Even though I am of Chinese descent, the waiters had assumed, because I speak perfect English, that I must not be interested in the authentic Chinese food.

Mathematical spaces—in the home and in the classroom—can be like this restaurant. Whom do we allow a peek at the secret mathematical menu? With whom do we share mathematical delights—puzzles, games, toys? Whom do we let into our information circle about mathematics—news, videos, social media posts? Whom do we shepherd toward doing more mathematics, and whom do we discourage? What conscious or unconscious assumptions are we making?

Akemi was a student of mine who did math research with me as an undergraduate. Her innovative paper linking game theory (the mathematical modeling of decision making) and phylogenetics (the study of relationships between organisms) was published in a highly regarded mathematical biology journal. She went to a top research university to pursue a PhD in mathematics, so I was surprised when I learned that Akemi had quit after one year.

She told me that she had had many negative experiences. Her advisor was never willing to meet with her, and she had faced uncomfortable experiences as a woman. She told me one example:

> At the beginning of the course, I consistently got 10/10 on my homework assignments, which were all graded by the TA. One day, Jeff [a mutual friend] told me that he was hanging out with our TA and someone asked the TA how the analysis class was doing. He went on and on about some "guy" named Akemi and how perfect "his" homeworks were and how clearly they were written, etc. Jeff told him I was a girl and the TA was shocked. (Jeff told me this story because he thought it was funny that someone both didn't know my sex from my name and reacted so dramatically to finding out.) After that, I never got remotely close to 10/10 on my assignments and my exams were equally harsh—most of the reasons for docked points were vague, with comments like "give more detail." I didn't feel like my understanding of the material diminished that quickly or dramatically, but I suppose it's possible that happened and I'm just misinterpreting the situation.

If you feel that something is not right with this picture, you are experiencing a telltale sign of flourishing: a desire for justice. Justice is a basic human desire.

Justice means treating people equitably—giving each person what they are due. It means speaking up for the powerless, who are most likely to be treated unjustly. Some religious traditions, including mine, advocate caring for orphans, widows, immigrants, and the poor. I see these categories in the mathematical community as well: those without an advocate, those without a vibrant mathematical family, those who are new to mathematics, and those without resources or access to mathematics. These are the powerless in mathematics.

Some speak of justice in two categories: *primary justice* and *rectifying justice.*[1] Both are important. Primary justice involves right relationships: treating each person with dignity and care, and establishing social practices and institutions that support these aspirations. We flourish when we treat others justly and when we are treated justly.

Rectifying justice is spotting something wrong and trying to make it right. If primary justice were commonplace, rectifying justice would not be needed. But injustices are everywhere. Coercive power relationships are destructive. People and institutions may subtly promote inequities without realizing it.

Simone Weil realized that correcting injustice must involve changing how we view others. "Every being cries out silently to be read"—to be judged—"differently." Akemi wished to be read fairly by her TA, but the TA may not even have realized what he was doing. This is the problem of *implicit bias:* unconscious stereotypes that subtly affect our decisions. Before we criticize Akemi's TA, we have to realize that the problem of reading others differently begins with ourselves. I've taken diagnostic exams that revealed my implicit bias; these helped me see in a compelling way how I am biased even though I try not to be.[2]

We all have implicit biases. Numerous experiments confirm results of the following kind: when given two nearly identical resumes except that one has a positively stereotyped name and one has a negatively stereotyped name (that depends on context, but often female or minority), judges will rate the positively stereotyped resume higher. This happens even if the judges come from the negatively stereotyped group. Similar studies confirm that math achievement is correlated with teacher and parent stereotypes. For instance, a 2018 study of elementary school math test performance showed that teachers scored girls lower (and boys higher) compared to outside evaluators who didn't know the identities of the students, and the long-term effects of this bias persisted downstream: those girls were less likely to sign up for advanced math courses in high school.[3] A 2019 study showed that the gender gap in math performance increases substantially when middle-school students are assigned to teachers with stronger gender stereotypes, leading girls to develop lower self-confidence, underperform, and self-select into less demanding high schools.[4] Parental attitudes and stereotypes compound the problem. These issues disproportionately affect women and minorities.

If you believe that mathematics is for human flourishing, you will be dismayed, if you look at the demographics of those in mathematical studies or careers, that we aren't helping all people flourish. Across all races and ethnicities, college students intend to major in STEM (science, technology, engineering, mathematics) at the same rates, but rates of STEM-degree completion for underrepresented minorities are little more than half those for other groups.[5] Low-income and first-generation college students complete college degrees at far lower rates overall than non-first-generation students, a challenge that also carries over to STEM disciplines.[6] Women drop out of PhD programs in STEM at higher rates than men. At the end of the STEM pipeline, the demographics are overwhelmingly white and male from higher socioeconomic backgrounds.[7] We are losing many with an interest in STEM, but the losses are more pronounced in marginalized groups—a deeply inequitable situation.

Even to talk percentages makes it sound like we live in a zero-sum world, as if one person going into a STEM profession means another person does not. Contrary to this perception, the world will need drastically more people with mathematical skills as we become ever more STEM reliant. The 2012 report *Engage to Excel* from the President's Council of Advisors on Science and Technology estimates that in order to maintain its preeminence in STEM, the US alone will need to produce an *additional* million STEM graduates over the next decade than it is currently expected to produce. That's not even emphasizing the obvious: we all lose when talent is overlooked, when we don't nurture those whose discoveries could benefit us all. A society cannot flourish if its people are not living up to their potential.

To rectify injustices, we have to talk about hard things that might separate us, like race, gender, sexual orientation, social class, the rural-urban divide, and the related ways in which some are marginalized in mathematics. Such conversations can bring up complicated emotions. We have to become more comfortable with talking about difficult topics, listening to one another's experiences, and recognizing the pain that's there. If you want to treat others with dignity and they're hurting, you don't ignore their pain—you ask: "What are you going through?"

It's not enough to say, "I don't think about that stuff—I treat everyone the same," because in any community, how any one member is doing affects the whole. Those of us in marginalized groups don't have the luxury of saying, "I don't think about that stuff," because that stuff affects us on a daily basis.

So let me encourage all of us to try having these conversations, to be quick to listen, slow to speak, and quick to forgive each other when we say something stupid. That'll happen if you start to have conversations, and we just have to show grace to one another if we make mistakes—it's better than not talking.

I'll share some of my experiences around race. I'm Chinese American. I grew up in a white and Latino part of Texas, and

I realized early on that my family had different customs than my friends—my clothes were different, the food in my lunch box was different—and that these things were isolating me. I was getting picked on all the time. I wanted to be white. I had few role models for being Asian American. I remember that my mother would always call me to the TV when the journalist Connie Chung was on, because it was so rare to see anyone who looked like us in the media. My dad would clip newspaper articles when any Asian Americans made the news. I was embarrassed to be Asian, so I tried hard to act white, even if I couldn't look white. This meant I would publicly deny anything Asian—not showing interest in Chinese food, not talking about Chinese customs—while I modulated my hair, clothes, and idioms of speech to be like those of my friends.

But in Chinese communities, I also don't fit in. I don't speak Chinese. I don't act Chinese. At Chinese restaurants, I'm viewed as white. That's why at authentic Chinese restaurants, I never get the secret menu. Asian Americans like me often feel like we live between two cultures, always viewed as outsiders, never fully fitting into either.

There are ways in which I benefited from being Asian. People assume I'm good at math because I'm of Asian descent. I never had anyone discourage me from taking a math class (as has happened to women friends of mine) or question whether I belonged at a math conference (as has happened to some of my African American friends). But even as I've benefited, I know that some of my Asian friends feel embarrassed when they don't live up to this stereotype.

The first time I didn't feel like a minority was when I moved to California, where there are so many Asian Americans. In Texas I would commonly get the well-meaning question "Your English is so good! Where are you from?" I'd say, "Texas," and inevitably, I'd hear, "No, I mean, where are you *really* from?" That happens less in California, and there's a feeling of freedom I have in not having to counter those verbal stings.

These days I'm used to being at math conferences and seeing a sea of white faces. So even I was a bit surprised that when I was elected as president of the Mathematical Association of America (MAA), a prominent blogger on race issues for Asian Americans, who goes by the alias Angry Asian Man, wrote a post about it.

Angry Asian Man looked at the photos of 100 years of past presidents on the MAA website, and given how many Asians he expected to find in math, he was astonished to note that they were all white except for me, and wrote a blog post sarcastically titled "Finally, an Asian Guy Who's Good at Math."[8]

I am the first president of color of the MAA. Minorities, including Asians, are easy to overlook when you think about who would make a good leader. This may not be intentional, but when we are asked to think about who is fit for this or that role, we often think of people just like those who have already been in office, and implicit bias creeps in. We don't realize what we might gain by having diverse people, new expertise, fresh ideas to draw from. The field of mathematics is itself poorer because of the voices that are not present. The math education professor Rochelle Gutiérrez reminds us that math needs a diversity of people in order to grow in new ways, not just that people need math: "The assumption is that certain people will gain from having mathematics in their lives, as opposed to the field of mathematics will gain from having these people in its field."[9]

I raise this discussion out of deep affection for all communities of math explorers. I want us to flourish, to make space to welcome new explorers from all different backgrounds, and we can do better.

Besides falling prey to implicit bias, there are other ways that mathematical communities misjudge people.

We assume that grades are a measure of mathematical promise. This is not a correct assumption, for many reasons. I used to worry that a student getting Bs in college math courses wouldn't be successful in graduate school. But I've seen many now who have gotten their PhDs and have flourished as mathematicians.

Grades are a measure of progress, but not a measure of promise. Everyone is in a different place in their mathematical knowledge. You see the snapshot, but you don't see the trajectory. You can't know how people will flourish in the future. But you can help them get where they want to go. When someone has trouble in mathematics, we should bolster our support, not lower our expectations.

It is easy to make assumptions about why people aren't performing mathematically, based on our own experiential vantage point. We can't imagine alternatives that we haven't experienced. We can't always know what personal issues someone is facing. A student once tearfully told me about the hours she spent filling out financial aid forms because her parents didn't speak English. Her immigrant family expected her to spend every weekend at home, and her home wasn't an environment conducive to homework. College was a culture shock, with many unwritten rules. This student was navigating many complex realities. Her performance wasn't the best she could offer, because of those realities.

Christopher Jackson has his own complex reality. Studying math in prison, he's mostly isolated from others. He hasn't been in school for a decade or more. He's developing his own way of learning and expressing mathematics. Expecting him to communicate math in the same way that I'm used to is unrealistic.

I'm sure that traditional assessments would not give a real picture of what he knows.

Low performance should not be used as an excuse to rob students of opportunity. The practice of tracking—sorting students into a dead-end course progression based on low performance—happens in some K–12 schools and is highly inequitable. Bias creeps into who is assigned to the "low" track. Those students are put into dead-end courses that do not prepare them for college and careers, are given access to less experienced teachers, and are assigned rote memorization exercises rather than rich meaning-making, sense-making activities. They cannot flourish mathematically. Tracking is a coercive practice that must be stopped.[10]

We assume that learning mathematics doesn't involve culture. This is a common assumption, especially if you are not part of a marginalized group. It leads to inaccurate assessments of student knowledge. A mathematician friend shared this example with me:

> On an examination, I asked the classical Fermi problem "Estimate how many piano tuners live in this city." A student timidly raised his hand. He whispered to me, "Is a piano tuner a device or a person?" Other students thought good piano players would tune their own pianos, like guitar players do. Some students thought piano tuners would work in a music store. Few students had a sense of how often a piano might need tuning, or how long it would take to tune a piano. This example opened my eyes to how important background experience can be in dealing with questions that may appear to be mathematical, but instead bring up all sorts of cultural or experiential issues.

I could have been one of those perplexed students, since a piano wasn't a household item for me. Now imagine a student without the requisite cultural experiences who constantly encounters obstacles like these. Would they feel like they belong? Cultural barriers are impossible to avoid, but if we are aware of them, we can mitigate their effects.

The math education professor William Tate points out that such experiences are common to African American children in mathematical spaces, who often encounter instruction based on white middle-class norms, and he contends that connecting pedagogy to the lived realities of African American students is essential for equitable instruction.[11] He advocates that teachers take a "centric" perspective: allowing and expecting students to center their problem solving in terms of their own cultural and community experiences, and encouraging students to think about how the same problem might be viewed from the perspectives of different members of the class, school, and society. For example, a teacher could reframe the problem about piano tuners as an estimation problem whose subject the students choose as relevant to their daily lives or struggles.

We assume that certain people won't be successful in mathematics, and we push them away from math. "You won't like the stuff on that menu." But if you believe that mathematics is for human flourishing, why would you do that?

In 2015, I had the great pleasure of running MSRI-UP (Mathematical Sciences Research Institute—Undergraduate Program), a summer research program for students from underrepresented backgrounds: Hispanic, African American, and first-generation college students. Later, I asked them to tell me about obstacles they'd faced in doing mathematics. One of them, who did wonderful work that summer, told me about her experience in an analysis course after she returned to her university:

> Even though the class was really hard, it was more difficult to receive the humiliations of the professor. He made us feel that we were not good enough to study math, and he even told us to change to another, "easier" profession.

As a result of this and other experiences, she switched her major to engineering.

Let me be clear: there is no good reason to tell someone that she shouldn't be doing mathematics. That's her decision—not yours. You may not know what she's capable of. One of my friends who is now a math professor described this incident that happened to him when he was a student:

> This faculty member had one of those private in-the-office conversations with me that begins with "I think it may be only a kindness to tell you that . . . ," followed by a stated concern that I was not really cut out for a career in mathematics. I've not done all that badly since then, and in fairness I have to add that the faculty member sought me out in later years to apologize for the comment. I consider the person a friend, but when I'm working with our graduate student training program, I do stress that any conversation that begins with "I think it only a kindness to tell you that" will almost never be a kindness.

Look at my friend now—a successful mathematician. It's too easy for such pronouncements to reflect personal biases and limited information.

Oscar, another student from MSRI-UP, told me about his experience as a math major who, unlike his peers and because of his background, did not enter college with any advanced placement credit:

> I noticed how different my trajectory was, however, while I was in a Complex Analysis course. A student was presenting a solution on the board which required a bit of a complicated derivation halfway through. They skipped over a number of steps, saying, "I don't think I need to go through the algebra . . . we all tested out of Calculus here anyway!," with my professor nodding in agreement and some students laughing. I quietly commented that Calculus was my first course here. My professor was genuinely surprised and said, "Wow, I did not know that! That's interesting." I was not sure whether to feel proud or embarrassed by the fact that I was not the "typical math student" that was successful from the beginning of their mathematical career. I felt a sense of pride in knowing that I was pursuing a math degree despite my starting point, but I could not help but feel as though I did not belong in that classroom to begin with.

The reason Oscar was in that class to begin with was the active support of another professor. Oscar said:

> She presented me with my first research opportunity and always encouraged me to study higher math. I was also able to confide in her about a lot of the internal struggles I had with being a minority in mathematics since, as a female, she had a similar experience herself! My complex analysis professor became one of my mentors as well. I think it was just an interesting moment because she didn't realize how her reaction to the situation could have hurt me (and I don't think she's necessarily at fault!). It was more that

her reaction piled onto the insecurities I held in regards to being a minority with a weak background in math.

Actually, Oscar didn't have a "weak" background—he had a standard background. I'm pleased to say that Oscar and his team from that summer have published a paper on their research, and he is now in graduate school.

You hear in Oscar's story the importance of having an advocate, someone who says, "I see you, and I think you can flourish in mathematics." Everyone can use this encouragement, but this can be especially important for marginalized groups who already have so many voices telling them they don't belong. Can you be that advocate?

We must be mindful to not set up structures for learning that disadvantage people with weaker backgrounds or make them feel out of place. When I was teaching at Harvard, there was a regular calculus class, an honors calculus class called Math 25, and on top of that—for those with very strong backgrounds—a super honors class called Math 55. Ironically, I regularly encountered students *in the honors track* who felt that they didn't belong in the math major, because they hadn't placed into the super honors track. I had to keep reassuring them that "background is not the same as ability." I wish that college and grad school admissions would remember this too. The mathematician Bill Velez says this about barriers at the graduate level: "In mathematics we value creativity, yet we evaluate students on knowledge. Departments erect barriers to keep down applications, and it works. Top-rated departments have few minority students."

Seeking justice can be a motivation to study mathematics, to rectify the injustices that exist in spaces of mathematical learning for the powerless in mathematics: the "orphans" who need advocates, the "widows" who need mathematical community, the "immigrants" new to mathematics, and the "poor" who face barriers of opportunity. Those who pursue justice in mathematics cultivate the virtues of *empathy for the marginalized* and *concern for the oppressed.* We sometimes don't recognize the constant burden of oppression that is experienced by the powerless until we start opening our eyes to see what they see. Those of us with power need to assist those who have no power.

The math teacher Josh Wilkerson engages his AP Statistics classes in a service-learning project, partnering with a homeless ministry in Austin, Texas, for survey research and data analysis. The students do a lot of "nonmath" reading on homelessness, to break down their assumptions on why people tend to become homeless. They also administer a survey and speak with a formerly homeless person face-to-face. As Josh puts it, "Hopefully, they recognize that behind every data point is a person, that person has a story, and that story is important."

Seeking justice builds in us a *willingness to challenge the status quo.* Many injustices are deeply embedded in the way institutions operate, whether that be school or workplace or home. People get different menus all the time, and no one says a thing. Long-standing inequities are hard to recognize because we operate within them and they're what we've always known. We need voices crying in the wilderness to call attention to the ways that we must change in order to treat each person with dignity and care in mathematical spaces.

I dream of that day when the secret menu will no longer be secret—when all people will be encouraged to develop their mathematical tastes so that someday they can be connoisseurs, even chefs, of mathematical cuisine.

RENTAL HARMONY

We've talked about just behavior in mathematical communities, but you can also use mathematics to study notions of justice. There's an area at the intersection of mathematics and economics called "fair division," which is concerned with how to divide things equitably among several people. A prototypical question is "How do you cut a cake fairly?" Math is involved in modeling people's preferences with sets or functions. I began doing research in this area when I encountered this problem:

> You and your college friends decide to rent a house together, and you have found a candidate house. However, the house has rooms of different sizes and with different features, and each of you has different preferences. Is it always possible to split the rent and price the rooms in such a way that each person will want a different room?

The answer is yes, under mild conditions—here's a result I proved in 1999:

> *Rental Harmony Theorem*
>
> Suppose the following conditions hold:
>
> 1. (Good house) In any rent division, each person finds some room acceptable at the proposed rent.
>
> 2. (Closed preferences) If a person prefers a certain room even as the rents change and approach a limiting rent division, then that person will continue to prefer that room in the limiting rent division.
>
> 3. (Miserly tenants) A person will always prefer a free room to a nonfree room.
>
> Then there exists a rent division in which each person will prefer a different room.

The proof involves ideas from geometry and combinatorics (the study of ways of counting things), and it yields a procedure for finding a fair division of rent. When a *New York Times* reporter used my procedure to solve his rent division problem, he wrote about it in an article and published an interactive app that implements this procedure. I encourage you to try the web app.[a]

By the way, if you drop the miserly tenants condition, the theorem is still true, but you'll need to allow negative rents; in other words, you can still find a solution, but you may need to pay someone to live with you!

a. The rental harmony result may be found in Francis E. Su, "Rental Harmony: Sperner's Lemma in Fair Division," *American Mathematical Monthly* 106 (1999): 930–42. The *New York Times* article is Albert Sun, "To Divide the Rent, Start with a Triangle," *New York Times,* April 28, 2014, https://www.nytimes.com/2014/04/29/science/to-divide-the-rent-start-with-a-triangle.html; the interactive web app can be found at https://www.nytimes.com/interactive/2014/science/rent-division-calculator.html.

September 5, 2018

I appreciate your efforts in trying to call up here for me; what everybody's doing on the outside is really starting to have an effect and I should be out soon. I got your letter on 8/31 (they had me in medical observation from 8/16 to 8/28 and I wasn't receiving any of my mail then). Hopefully, I'll be able to send you this in an e-mail (I've been on a hunger strike 24 days, and now I've seen my unit manager 3 times in the last 5 days—before then I hadn't seen him in a month and half) because it seems me and the administration have reached a resolution and this should be over very soon. I'm supposedly going to find out tomorrow.

One of the important things I'm learning from you is that mathematics is around and about ideas. Not necessarily just the ideas inside of mathematics but also the ideas parallel to mathematics. (I'm about to digress.) I remember my first incursion into the realm of ideas. . . . I was 16 and I was in a group home; my caseworker at the time gave me 3 books, [including] *The Art of War* by Sun Tzu, great book, even though the title might throw some people off, on a philosophical level or even a metaphysical level for dealing with strife, whether it be internal, external, personal, interpersonal or whatever. . . . These books ignited my interest in philosophy, which then went on to politics, economics, business, and then eventually back to mathematics.

(9/9/18: I came off hunger strike 9/7/18. I'm designated to another institution—I should be on a bus leaving out of here this coming week. I started this letter while I was still not eating, because I had an idea of what I wanted to write, but it was getting difficult because my glucose levels were averaging around 65.)

And through every new idea that I've encountered, whether they be ideas within a discipline or ideas across disciplines, I started to notice principal ideas: order, relation, organization, structure, process. . . .

And from what I'm gleaning from you about mathematics being around and about ideas, I have to ask, do you think mathematics unites these ideas?

Chris

11
freedom

Any teacher can take a child to the classroom, but not every teacher can make him learn. He will not unless he feels that liberty is his, whether he is busy or at rest; he must feel the flush of victory and the heart-sinking of disappointment before he takes with a will the tasks distasteful to him and resolves to dance his way bravely through a dull routine of textbooks.
Helen Keller

Freedom makes a huge requirement of every human being. With freedom comes responsibility.
Eleanor Roosevelt

I thought I had chosen the right story for the occasion. Assembled before me was a group of eager Latino and African American children from an impoverished Los Angeles neighborhood. On this Saturday morning, I was serving with a program in which volunteers read books to kids. I had selected a delightfully illustrated picture book about going to the beach, and I thought it would be received well. But after reading a few pages in a most spirited voice, I could tell that the kids were not sharing my enthusiasm.

I paused, and asked: "How many of you have ever been to the beach?"

To my surprise—though this part of L.A. is just fifteen miles from the ocean—only one of the eight children raised a hand. Wasn't going to the beach a quintessentially Californian thing to do?

Upon reflection, I realized that in a low-income neighborhood, parents often work multiple jobs to make ends meet, so they may not have the time or the resources to drive to the coast. And when an African American friend of mine heard this story, he explained how African Americans were systematically excluded from beaches and swimming pools because of Jim Crow segregation, not only in the South but all over the US, including Los Angeles. I was completely unaware of this.

Alas, I had missed important historical, cultural, and economic contexts that made the beach inaccessible to these kids. It made me reflect on how I motivate my students to pursue mathematics. What contexts am I missing that I should be more aware of? What are the primary experiences that have shaped or are shaping them, and do those present obstacles to or opportunities for learning math? What are the unique strengths they bring to mathematical pursuits? And in what ways do mathematical spaces, like beaches, say "Open to all" but still feel restricted?

For me, the beach became a metaphor for various freedoms

that are hallmarks of doing mathematics—freedoms delivered to some and denied to others. Just as they should be manifest at every beach, the freedoms we'll discuss should be present in every mathematical space. They are part of the allure of doing mathematics for those fortunate enough to experience math as it should be experienced. Conversely, the denial of those freedoms contributes to the fear and anxiety that many people feel toward math.

Freedom is a basic human desire. It is a central idea behind historic human rights movements and a sign of human flourishing. We seek freedom in big ways—think of the Four Freedoms that President Franklin D. Roosevelt said all people should have: freedom of speech, freedom of religion, freedom from want, and freedom from fear. We also seek freedom in small ways that can feel just as important, such as freedom with our time or freedom to make our own decisions.

I want to highlight five freedoms that are central to doing mathematics: the freedom of knowledge, the freedom to explore, the freedom of understanding, the freedom to imagine, and the freedom of welcome. As a math explorer, you should be aware of these freedoms so that you can claim them for yourself and aspire to fulfill them for everyone you encounter.

The freedom of knowledge is easy to underestimate, because if you have this freedom, you take it for granted, and if you don't have it, you are completely unaware of what you're missing. You have to know about the beach and know its many options for recreation—how to swim, surf, dive, tan, picnic, play volleyball, etc.—if you are to experience its freedom. These seem obvious to anyone who's been there, but if you are like the kids who didn't know about the beach, either because you were never told or because someone prevented you from going, you will not know the joys that await there.

Within mathematics, the freedom of knowledge is also fundamental. If you know just one method for attacking problems, you are limited, because that method may not work well for your particular problem. But if you have several strategies, you have the freedom to choose the option that is the simplest or most enlightening. Mathematics equips you to look for multiple ways to solve problems.

The mathematician Art Benjamin is a human calculator—he can multiply five-digit numbers in his head. While this sounds impressive, the mathematical fun for him is not in the calculation. The fun is in thinking of multiple strategies for doing a calculation easily and choosing the one that works best.[1] I'm not as practiced as he is, but I also rely on such skills to do calculations. For instance, if I want to multiply 33×27 in my head, I can think of four different ways to do it.

I can do it the "standard" way, which means taking thirty 27s and three 27s and adding them. That's $(30 \times 27) + (3 \times 27) = 810 + 81 = 891$. I don't find it so easy here to hold all the intermediate calculations in my head.

Or I can do it by factoring 27 as 3×9 and first multiplying 33 by 3, then multiplying that product by 9. That's $(33 \times 3) \times 9$, which is $99 \times 9 = (100 \times 9) - (1 \times 9) = 900 - 9 = 891$. This seems easier than the standard way.

Or I can factor 33 as 3×11 and first multiply 27 by 3 (which is 81) and then multiply that product by 11. That's 81×11, which is easy if I know the shortcut for multiplying by 11: take the digits 8 and 1 and insert their sum, 9, in between to get 891.[2]

Or I can see that this algebra identity might be helpful: $(x - y)(x + y) = x^2 - y^2$. So if I recognize $27 = 30 - 3$ and $33 = 30 + 3$, then the desired product of 27×33 is just $30^2 - 3^2 = 900 - 9 = 891$.

If someone asks me to do this calculation quickly, I will look at the quiver of arrows I have and choose the best arrow to attack this problem. For me, that would be the last way. The freedom of knowledge gives us a large quiver.

My thinking about the freedom of knowledge was inspired by Christopher Jackson. He said to me once that *freedom is knowing all the options available to you.* He came to understand this while playing chess in prison. Another player can dominate you and control you by limiting your options on the chessboard, as he noted:

> It is [a sign] of a skilled chess player that he or she can play well out of any position or circumstance on the board. The person who is unaware of his or her options is like a player who is in a bad position. Because even if there are fruitful paths available to you that you are not aware of, they may as well not exist. This is like if a player finds himself with 2 bishops against a lone king, but doesn't know that 2 bishops can checkmate, he or she will call a stalemate. But when this person is shown (educated) that 2 bishops can checkmate, they will always recognize this situation as a win. And this is in my opinion the main bridge of education, to take people to a place where they can recognize the pathways to success. . . .
>
> By "climbing our visions of ourselves," education allows us to transcend ourselves, and thereby help others to do the same.

The freedom of knowledge speaks to the larger role of education for all of us: to bring us to a place where we can recognize pathways to flourishing.

A second basic freedom that should be present in mathematical learning is *the freedom to explore.* Just like the wide expansiveness of the beach—with its shells, its sounds, and the treasure we fancy buried underneath—the learning of mathematics should be a place for exploration, so as to stimulate creativity, imagination, and enchantment. But some teaching styles don't offer this kind of freedom. I think of the difference between how my mom and my dad taught me math, a study in contrasts between obligation and exploration.

My parents wanted me to learn math at a young age, so even before I went to school, my father would teach me numbers and arithmetic. Since he was busy with his own work, he would make up long worksheets of addition problems to keep me occupied. I did them as a dutiful kid, but didn't find them very much fun. "Do this one over," he would say. "You can't go out to play until you get every one right."

My father's approach was a one-way transmission of information. He showed me what to do, but left me to myself to do worksheets. I followed the rules he taught me for arithmetic, often without understanding. I learned how to add numbers bigger than ten by "carrying," but I didn't have any idea what I was doing. I was following recipes. And my father's praise and rewards were always connected to my performance. Now, my dad, in all fairness, was a good dad, but in an immigrant Asian American family, I could be shamed for turning in anything less than a perfect paper. That's not freedom.

By contrast, my mother's approach was relational. We played games that encouraged numerical thinking and pattern recognition. She sat with me, and together we read books about counting. And the books we read were also relational—full of wonder and delight. They invited more questions. Like: Why does that Dr. Seuss character have eleven fingers? It's not even five fingers on one hand and six fingers on the other, as you might expect, but rather four and seven! This fanciful strangeness invited further imagination. With my mother, I had the freedom to explore, and the freedom to ask questions, the freedom to think ridiculous thoughts. Questions and fanciful thoughts were praised.

This freedom is also at the heart of mathematics at higher levels of learning. As a high school senior, I attended a lecture for prospective students at the University of Texas at Austin. The topic was infinity, and the speaker was the math professor Michael Starbird. His lecture style was different than anything I had experienced in high school. It was highly interactive, and he was constantly asking questions of the audience—as if to invite us to be explorers together. I'd never before been in a room with three hundred people where everyone was engaged and paying attention. This kind of interaction exemplifies a teaching style known as *active learning.* I left that lecture thinking: Wow, if every class here is like that, college is going to be a lot of fun.

So I enrolled at Texas. Having placed out of calculus, and thinking I was "good" at math, I jumped into the subsequent course. It was in a traditional lecture style—this meant that the professor lectured without much interaction, and we took notes. On the first day, he began talking about matrices, a topic that I had never seen before and that wasn't on the list of prerequisites for the course. (A matrix is an array of numbers, and usually discussed in the class *after* this one.) And then he started *exponentiating* matrices, which means he took the number *e* and wrote an array of numbers as an exponent. For me, that was category confusion: like asking me to use an avocado to brush my teeth, or putting my cat in my wallet.

I looked around and assumed that everyone else knew what was going on. I was intimidated, fearful of asking any questions because no one else was and the professor wasn't inviting them. Symbols were flying by as if a broken keyboard were stuck on the Symbol font. I dutifully took notes, but I had no idea what I was writing down. And that was just the first day of class. All semester I struggled to keep up, my understanding always two weeks behind. That was not fast enough to help me on my homeworks and exams, where I was often guessing at solutions I didn't understand. I was a hamster on a wheel powered by someone else, fearful that any mistake would mean I'd tumble off and do poorly in my first math class in college. This was not freedom.

This story highlights a third kind of freedom that mathematics offers: *the freedom of understanding*. I was learning that if you go through life pretending you understand, you will always be a slave to the things you don't understand. You will continue to feel like an impostor, believing that everyone else knows what is going on and you're the one who doesn't belong. By contrast, true understanding means you have to devote fewer brain cells to remembering formulas and procedures, because everything fits together meaningfully. Math education should promote, rather than inhibit, this freedom, but as learners we must strive for deep understanding even when our education isn't promoting it. This is where the hard work is.

After that first course, I almost didn't become a math major. But I decided to give it one more try. I took a course with a professor who was much more interactive and approachable, and I began to feel more confident again. Then, the next year, I took a class with Starbird. The class topic was topology—the mathematics of stretching things. Or, a little more accurately, it's the study of properties of geometric objects that don't change when you continuously deform the objects. For that reason it is sometimes called "rubber-sheet geometry." This meant that drawing pictures was very important in this course, while numbers were almost nonexistent!

To my delight, Starbird was teaching in an "inquiry-based

learning" format. There were no lectures. Instead, we were given a list of theorems and provided the challenge of discovering their proofs for ourselves. Through guided interaction with him and with one another, we learned how to present our ideas and subject them to constructive scrutiny by peers. But the underlying strength of the course was how the professor used this format to encourage a different classroom culture. He created an environment where questions were praised and unusual ideas were welcome. He was giving us the freedom to explore.

Relationships with each other were central to our explorations. We learned in this environment how to proclaim, "My proof is wrong," without shame or judgment. Indeed, a wrong proof was always a point of delight, because it meant we were seeing something subtle, and it was a springboard to further investigation.

I've seen professors foster this kind of culture in more traditional lecture formats too, using active learning methods. In such classes, every day can be like a Dr. Seuss poem, filled with surprise and wonder, where the fanciful is celebrated.

A fourth freedom present in mathematics is *the freedom to imagine.* If exploration is searching for what's already there, imagination is constructing ideas that are new, or at least new to you. Every child who has ever built a sand castle on the beach knows the limitless potential of a bucket of sand. Similarly, Georg Cantor, whose groundbreaking work in the late nineteenth century gave us the first clear picture of the nature of infinity, said, "The *essence* of *mathematics* lies precisely in its *freedom.*"[3] He was saying that, unlike in the sciences, the subjects of study in mathematics are not necessarily tied to particular physical objects, and therefore mathematicians are not constrained like other scientists in what they can study. A math explorer can use her imagination to build any mathematical castle she wants.

My topology class taught the practice of imagination. Topology, as I mentioned earlier, is the study of properties of geometric objects that do not change when you stretch the objects in a continuous way. If I take an object and deform it without introducing or taking away "holes," I haven't changed it topologically. So a football and a basketball are the same in topology, because one shape can be deformed into the other. On the other hand, a doughnut is not the same as a football in topology, because you can't turn a football into a doughnut without poking a hole in it.

Topology is an entertaining subject because we get to construct all sorts of groovy shapes by cutting things apart, gluing things together, or stretching things in weird ways. We often envision moving around inside these shapes, so we call them *spaces.* Topology fans have lots of fun imagining their own weird spaces, usually in response to some fanciful question, like "Does there exist an object with this-or-that pathology?" (Yes, we use the word *pathology* in math, just like in medicine, to describe anything strange or abnormal.) Then one uses one's mind to conjure up an example. For instance, there's the Lakes of Wada: three connected regions ("lakes") that can be drawn on a map so that they have exactly the same boundary—any point on the edge of one lake must be on the edge of all three. The construction was named playfully after the mathematician Takeo Wada, who invented them. Then there's the Hawaiian earring, an ornate object with infinitely many, successively smaller, rings, all touching at one point.[4]

A rather famous example of a pathological space (among mathematicians, anyway) is the Alexander horned sphere. A sphere is a bubble-shaped surface. The space outside a perfectly

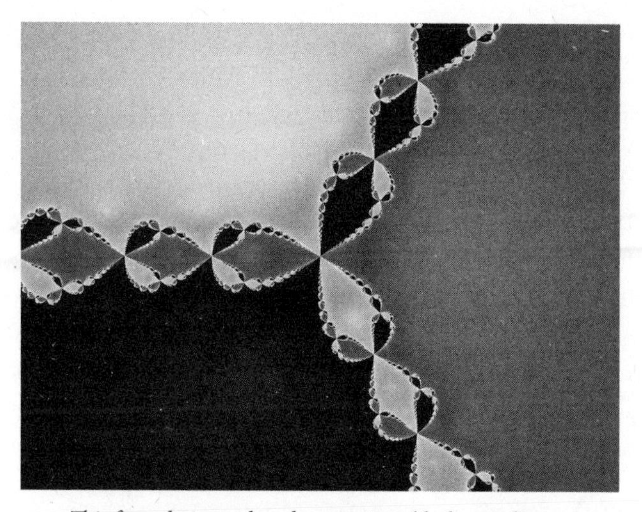

This fractal picture has three regions (dark-, medium-, and light-shaded "lakes") that share the same boundary, though unlike the original Lakes of Wada, each lake here consists of disconnected basins.

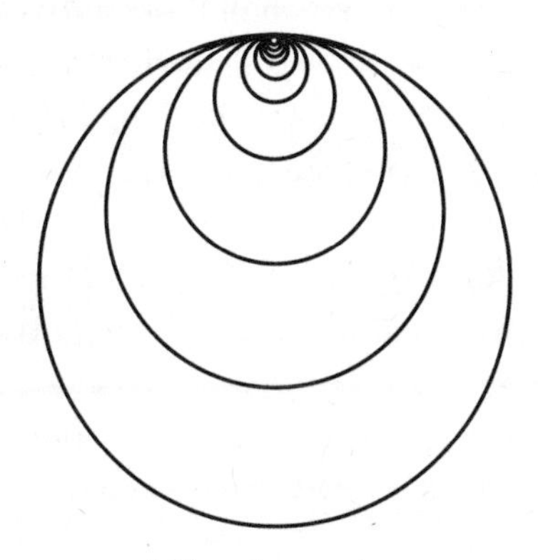

A Hawaiian earring.

round sphere has the property that it is "simply connected," which means, roughly speaking, that if you take a string outside the sphere and tie its ends together to form a loop, you won't be able to tie a loop that will get caught on the sphere—it can always be pulled off and separated from the sphere. (This contrasts with a donut, whose exterior space is not simply connected: if you run one end of a string through the hole in the center of the donut and then tie the ends together, you will *not* be able to untangle it and separate it from the donut.) The question that J. W. Alexander was thinking about in 1924 when he imagined his horned sphere was, Is it possible to deform a bubble in such a strange way that two distinct points of the bubble never touch, but the space outside the bubble is not simply connected?

Alexander initially thought that the outside of any deformed bubble must be simply connected.[5] But then he produced an example where the outside is not! His imaginative construction can be described as follows (this is not exactly his construction, but it's topologically the same): Take a bubble and push out two "horns." Now from each of those horns push out a pair of pinching fingers in such a way that they almost interlock with the other pair of pinching fingers. Because the pinching fingers don't quite touch, you can repeat the process on a smaller scale, pushing out a pair of tiny pinching fingers—which interlock but do not quite touch—from each of the preceding set of fingers. Keep doing this, and in the limit, you have the Alexander horned sphere.

A loop of string encircling the base of one of the original horns cannot be disentangled and separated from the horned sphere, precisely because of the limiting process of the interlocking pinchers. If the pinchers ended at some stage without taking a limit, then the loop could easily slip off. This surpris-

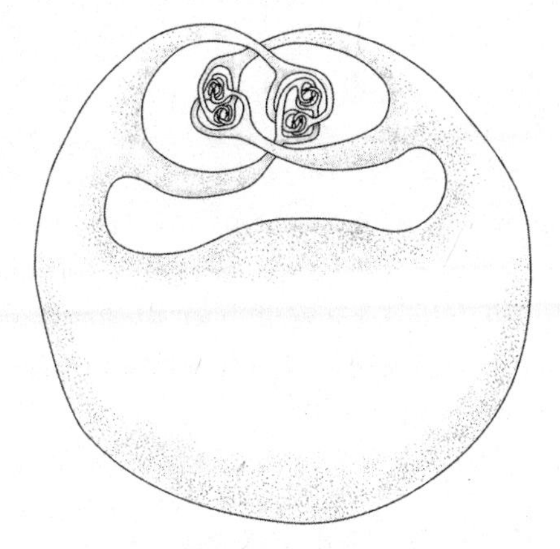

The Alexander horned sphere.

ing construction requires imagination not only to think of it, but also to verify that in the limit the horned sphere is, indeed, still a sphere. You can imagine zooming in and seeing the fractal nature of successive levels of pinching horns: at every level of detail, the view looks the same.

The freedom to imagine infuses mathematics with a dreamlike quality. Make a wish, and *voilà!* Your dream comes alive.

How much more fun could mathematical learning be if, at every stage, we had opportunities to use our imagination? You don't need to be doing advanced mathematics to do this. In arithmetic, we can try to construct numbers with fanciful properties. What's the smallest number divisible by all the digits in your date of birth? Can you find ten numbers in a row that are not prime? In geometry, we can design our own patterns and explore their geometric natures. What kinds of symmetry exist in the patterns you like? In statistics, we can take a data set and come up with creative ways to visualize it. Which ones have the best features? If you're learning mathematics from a dull textbook, see if you can modify the questions so that they increase your imaginative capabilities. In doing so, you are exercising your freedom to imagine.

Unfortunately, the prior freedoms—the freedoms of knowledge and of understanding, the freedoms to explore and to imagine—are difficult to secure without the last freedom, which is *the freedom of welcome.* This is a freedom missing from many mathematical communities.

Beaches, as I learned, have a historical association that is exclusionary, which keeps people, even today, from enjoying those spaces. Imagine this scenario at the beach. There's no longer a sign saying that you aren't allowed, but you don't come very much, because your parents never came at all. There's no one chasing you out, but you get sideways stares. People question whether you meant to go to *another* beach. Some think you're the service staff at the beach showers and ask for more paper towels in the restroom. Others avert their gaze and clutch their children tightly when you walk past. People make up seemingly arbitrary rules for you, telling you that you can't cook *that* food for your picnic, or play *that* game on this beach. You go instead to the volleyball courts for a pickup game, but no one invites you to play. They don't expect that you know or will want to learn the game. The beach may be open to you, but you aren't really welcome.

Sadly, mathematical communities can be like that. We say we value diversity, but there are exclusionary undercurrents. Consider these examples.

Your name is Alejandra, and you've noticed that in every math textbook since grade school, the names in generic examples are all white male names. In middle school, you come up with novel ways to solve problems, but your teacher never seems interested in solutions other than the ones she knows. Your high school math teacher is lecturing, and he makes eye contact only with boys.

You place into an advanced math class in college and find the work challenging, but the professor encourages you to drop to a less advanced class rather than encouraging you to continue. You're a college athlete with a demanding practice schedule every afternoon, but the professor makes himself available only for afternoon appointments. A professor calls a proof "trivial" and "obvious"; you think there's something wrong with you because it isn't trivial and obvious to you. He notes that "students like you don't often do well in this major." A math competition is going to take place; the organizer invites every math student but you to the practices. You come from a culture that prizes community and storytelling, yet your math professors talk about math as if it's completely devoid of any history or culture, and all the assigned work is to be done alone.

You decide to go to graduate school in math, but there are few women in the program and no Latina students like yourself, and certainly no Latina women on the faculty. No one knows how to pronounce your name; they call you "Alex" without your permission. The graduate student lounge in your department has no art or plants or color; it feels sterile, and you certainly don't want to hang out there. The other students seem very competitive, and quick to point out others' mathematical errors in unsupportive ways. Your advisor seems uninterested in your life outside work, even when you signal that you're struggling with child care. Yes, you've made a decision to start a family in grad school, but the administration seems inflexible in handling that.

You become a mathematician, and are thrilled to get a job at a college where teaching is valued, but your friends at research universities ask with pity, "Are you happy there?" When you go to conferences, your small stature and dark complexion mean that you are often mistaken for "the help" at conference hotels. When you publish papers collaboratively, people always think that the other person did more of the work. So you feel pressure to publish papers alone. You love all the things that mathematics offers, but it doesn't feel like it's worth *this.*

Taken together, Alejandra's experiences can feel life draining and oppressive, even though the people involved may have had the best intentions and been completely unaware of what she was going through. Collectively, they are a coercive use of power. Alejandra does not have the freedom of welcome. You might wonder why she hangs in there at all.

To be welcoming means more than just allowing people to coexist. It means extending an invitation of welcome—to say, "You belong," and follow it up with supportive actions. It means maintaining high expectations and providing high support.

Expectations can influence how a student does in class. There is substantial research on "expectancy effects," which shows that teacher expectations can affect how students learn. The most famous is the 1966 Rosenthal-Jacobson study, in which students were given a fake aptitude test and their *teachers* were told which students were expected to "bloom" (when in reality the so-called good students were randomly selected). Over the next year, those students did better than their classmates.[6]

This is a silent captivity of expectations. It holds both student and teacher captive. Teachers are bound by a limited imagination of a student's potential. Students are bound to someone else's idea of who they can be, and they don't have the freedom to be free. A freedom of welcome would say, "I believe you can succeed, and I will help you get there."

In the book *Teaching to Transgress,* bell hooks discusses her experience as a student in an all-black school in segregated

America. She praises the teachers who were on a mission to help the students reach their highest potential.

> To fulfill that mission, my teachers made sure they "knew" us. They knew our parents, our economic status, where we worshipped, what our homes were like, and how we were treated in the family. . . .
>
> Attending school then was sheer joy. I loved being a student. I loved learning. . . . To be changed by ideas was pure pleasure. . . . I could . . . , through ideas, reinvent myself.[7]

You can hear how those teachers practiced the freedom of welcome. They got to know everything about their kids, not just their academic performance. These students' education was rooted in community. Because of the freedom of welcome, hooks had other freedoms: the freedom to explore ideas and the freedom to imagine a new identity for herself.

By contrast, after schools were integrated and she changed schools,

> knowledge was suddenly about information only. It had no relation to how one lived, behaved. . . . We soon learned that obedience, and not a zealous will to learn, was what was expected of us. . . .
>
> For black children, education was no longer about the practice of freedom. Realizing this, I lost my love of school.[8]

The beach was now open, but there was no welcome, no community or hospitality. hooks was captive to expectations: always feeling like she had to prove herself. She was afraid that if she spoke up, she would be perceived as stepping out of bounds. Education felt like domination. Without the freedom of welcome, she lost all the other freedoms.

To be clear: I'm not advocating segregation. I'm saying that real welcome must involve real freedom, especially when that freedom has been denied in the past.

These freedoms in mathematics are associated with several virtues. The freedom of knowledge leads to the virtue of *resourcefulness.* We can take the tools we know about and bend them to solve our problem. Having the freedom to explore builds *fearlessness in asking questions* and *independent thinking,* when we are not shamed for brainstorming aloud and we experience the joy of discovery. It also builds in us the skill of *seeing setbacks as springboards,* as we learn to not simply discard wrong ideas but explore how they can lead us to good answers or push us off into new areas of investigation. The freedom of understanding builds our *confidence in knowledge,* because understanding builds a firm foundation of facts cemented together by meaning and insight. And the freedom to imagine encourages the virtues of *inventiveness* and of *joyfulness,* because that freedom gives you room to explore and to take delight in all the wild things your mind might conjure up.

When I encountered difficulties in graduate school, feeling underprepared and out of place, with professors questioning my ability to succeed, these virtues rescued me. I knew I had the beginnings of research skills, because I'd experienced independent thinking. I leaned on my fearlessness in asking questions to speak up when I didn't understand. I knew the joyfulness in creating mathematics for myself. And the confidence in the knowledge I already had helped me to trust that I would eventually catch up, through earnest effort and hard work. I could, borrowing Helen Keller's words, resolve to dance my way bravely through when I felt the freedoms that I had.

Freedom is a crucial ingredient of learning and doing mathematics, so we ought to consider what freedom entails. I know that some people define *freedom* as "the absence of constraints," as if it means "do whatever you want." I don't believe that's what true freedom is.

True freedom never comes without cost, relationship, or responsibility. Think of that teacher who poured their time and energy into you, gave you the space to ask questions, and showed you how to explore the beach and imagine the castles you could build. Think of that parent who provided whatever resources they had so that you could get to the beach, and instilled in you a resolve to overcome any obstacles. Think of the real welcome that some provided, helping you to enter, the welcome that took real energy and care. And think of the cost of your own commitment to invest yourself in a subject that is deep and wide and beautiful, so that you now have the freedom to flourish on that beach in ways you never could have otherwise.

Those of us who have experienced the freedoms of mathematics have a significant responsibility to welcome others to those freedoms as well.

UNKNOWN POLYNOMIAL

This problem may seem to require a little more background than other puzzles in this book, but if you stick with it, you may enjoy its solution. A *polynomial* is an algebraic expression like $x^3 + 2x + 7$, and you can evaluate it at a number by plugging in that number for x. So the value of $x^3 + 2x + 7$ at -1 would be $(-1)^3 + 2(-1) + 7$, which equals 4. The *degree* (highest power) of this polynomial is 3. The numerical factors involved in each term are called *coefficients* (here, 1 is the coefficient of x^3, 2 is the coefficient of x, and 7 is the constant coefficient).

Here's the puzzle:

Suppose I have a polynomial with nonnegative whole number coefficients. You don't know its degree. Your goal is to determine this polynomial, but you are allowed to ask me questions only of the following form (where k is a specific whole number):

"What is the value of the polynomial at k?"

What is the fewest number of questions needed to determine my polynomial?

I love this problem because it seems like you don't have enough information. But there's a lot of freedom in the questions you can ask. Sam Vandervelde first shared this delightful problem with me.

January 28, 2018

I've transferred now 2 times after my security level dropped. In my last institution I met some guys that I bonded with well—we were rowing in the same direction. I learned a few things from them. I'm still a fairly young guy: I'm 31 now and I've been in jail since I was 19, so there's a lot for me to learn. Among the 7.1 billion of us on earth there is a vast commonality, but even among that commonality, there is a (small) infinity of differences and uniquenesses. As of now I haven't found one person here even close to the wavelength that I'm on. But even when that sameness of vibration can't be achieved I realize you have to be open to community and planting seeds and being planted on, because there is something that can be learned from everybody. But as of right now I just have a couple of people I talk to most days. I left 3 friends at the institution I just left and 2 when I left Kentucky.

Chris

12
community

The real satisfaction from mathematics is in learning from others and sharing with others. All of us have clear understanding of a few things and murky concepts of many more.
Bill Thurston

What is belonging? The extent to which an individual feels accepted, valued, and legitimate within the community.
Deanna Haunsperger

Ricardo Gutierrez is a New York City native, the son of immigrants in a working-class neighborhood. His father never finished high school, and his mother didn't make it past the

eighth grade. He wrote to me in 2017 after reading a speech I gave about flourishing in mathematics. He had shown an aptitude for math at an early age, but lacked a mentor to guide him. So he pursued another interest in college, and for the past nineteen years has had a successful career as an audio engineer—not really a technical job, but, as he puts it, "I work on music and make it sound better." He loves his career—in fact, one of the projects he worked on was up for a Grammy—but he felt that something was missing:

> I just yearn for more, and it felt like there was a hole in my life. I yearn for more knowledge and more thinking about the types of things that mathematics and computer science gives me. Maybe it is the drilling down on logical problems that I find appealing, even meditative. So maybe it is that I have outgrown the boundaries of my career and I wanted to supplement it with mathematics and programming. . . . Maybe it is more appropriate to say that having mastered my career, it has become rote.

He took the brave step of returning to school at age forty, in a program for nontraditional students. He says:

> While the rigor and pressure of being in a really tough academic environment have been almost impossibly difficult—especially after I've been out of practice for so long—it is the entrenched feeling that I don't belong in those Math and CS classes that has sometimes been the most harmful. Those feelings are most probably tied to my early life and the fact that any dreams I may have had then were in discord with the cold realities of my neighborhood and life, and that I didn't have any mentor then to disabuse me of that great untruth. That distorted reality, the "I'm

not supposed to be here," runs as an infinite loop process in even unseen ways. It is a constant struggle.

This feeling—*I don't belong*—can be quite crippling. And this is where community is terribly important: for us to feel belonging. Because if, as Parker Palmer says, "To teach is to create a space where the community of truth is practiced," then we are called to speak truths that others cannot see for themselves.[1] We can reassure them that they do belong.

None of us can flourish without a supportive community—people with whom we can share joys and sorrows, hopes and fears. A community helps us normalize struggle, and realize "I'm not alone in my struggle."

Community is a deep human desire. Thus, it serves as an entryway for many people to flourish in mathematical pursuits, whether recreational, educational, professional, or in the home. By "mathematical community," I mean any group of people who gather together over common mathematical experiences. You are forming a mathematical community in your home when you share math jokes, show enthusiasm for mathematics, make geometric objects, read stories together that involve math, or cook together—adding fractions for a recipe and talking about it. You are entering a mathematical community when you walk into a math class or if you join a game of strategic thinking.

Community is not a word that most people associate with mathematics. Instead, a popular view of a mathematician is a lone person toiling for years on a problem in solitude. Indeed, several recent solvers of famous problems match this description. In 1993, Andrew Wiles announced a (then slightly flawed) proof of the 350-year-old Fermat's last theorem, which is tantalizingly simple to state: the equation $x^n + y^n = z^n$ has no solution in whole numbers for $n > 2$. Wiles toiled in secret for seven years on this problem.[2] In 2003, Grigori Perelman proved the century-old Poincaré conjecture in topology, which, roughly speaking, says that a bounded three-dimensional object without any holes must be spherical. No one knew that he was working on it.[3] And in 2013, when Yitang Zhang proved that gaps between prime numbers are bounded—a major breakthrough in attacking the famous twin primes conjecture—no one in the field knew who he was.[4] These examples, which made news for their exceptionality, feed the myth that math should be a solitary endeavor.

In reality, mathematics has always been collaborative, because people gather together over many activities that are mathematical—learning, reading, games, research. We spend time in communities to enjoy math because, as Bill Thurston said (in response to a person who worried that they would never do anything original in math), the real satisfaction is in the learning and sharing.

Professionally, mathematics is becoming even more collaborative than it was in the past. A 2002 study showed that the proportion of authors doing joint research in math was 81 percent in the 1990s, up from 28 percent in the 1940s.[5] In 2009, the mathematician Tim Gowers famously enlisted the Internet to collaborate to find an elementary proof of the Hales-Jewett theorem, a result that roughly speaking asserts that versions of tic-tac-toe with many more dimensions and any number of players must always have a winner. A growing number of classroom math teachers are employing active learning techniques, which use class time for student participation and collaboration. With the rise of social media, math teachers are also connecting in novel ways to share ideas and form new interest groups. Working in teams is central to the way that math explorers interact today and an important skill for careers in business, industry, or government.

Community serves an important function in bringing people together in mathematical exploration—helping them to grow in the virtues promoted by socialization. The most engaging math-themed programs center on fostering community among participants, and you'll find a group to serve almost any niche, from kids to teachers to researchers.[6]

However, building community should be more than just bringing people together over mathematics. We must be vigilant to overcome obstacles that are sometimes more prevalent in mathematical communities.

Mathematical communities are often too focused on achievement—frequently, a narrow kind of achievement. Hierarchies get reinforced when we rank people according to a singular "ability." We do this mentally when we evaluate who is "good at math." We often signal to others that there's only one way to be successful in mathematics—by forcing kids to do math quickly, or rushing students into calculus in high school, or telling professionals that they aren't "real mathematicians" if they don't do research. In reality, there are multiple ways to be successful. Mathematical achievement is not one-dimensional, and we must stop treating it like it is. Too often we think of math like a pole in the ground: there's only one way for a vine to grow up the pole. In reality, math is like a trellis: as a vine, you can find your way up at multiple places where the trellis meets the ground, and you can grow in multiple directions along the trellis.

Thus, those who desire community in mathematics must develop strategies for combating the one-dimensional view of math. In the classroom or the home, we can praise others for showing the virtues cultivated by mathematics and remind them that these are part of mathematics too. Persistence, curiosity, habits of generalization, disposition toward beauty, thirst for deep investigation—and the many other virtues I've discussed in this book—are all ways to evidence growth in mathematics. In high schools and colleges we can develop multiple pathways into mathematics rather than funneling everyone through calculus. We can create math clubs based on enjoyment rather than math clubs based on elite performance. At the professional level, we can value the diverse ways that math teachers and researchers contribute to advancing mathematical understanding. We can hold up a wide array of role models in an exciting collection of mathematical careers.[7]

Mathematical communities can be very hierarchical, even if they don't wish to be. At my local hiking club, there's bonding over a shared love of hiking. When it's time to go, we split into different skill levels to do the hike at various speeds. I have no trouble admitting that I'm a slow hiker, and I feel no shame in joining the beginners' group. The joys of hiking—the scenery, the camaraderie, the quiet reflection—are separate from the skill of hiking. Similarly, if I attend a piano concert or a baseball game, the joy of watching is separate from the skill of playing.

By contrast, in mathematics the joy often requires the skill. For instance, if I'm going to a math lecture, it's not exciting to me unless I understand what's going on. The joy of the talk is not just knowing the theorem statements, but also following the proofs. Unfortunately, proofs are hard to absorb in a short time frame, and speakers are often not challenged to make their talks accessible to their desired audiences. By now, I've gotten used to the discomfort of not understanding everything, and I know it's normal, but it's easy for a newcomer to feel out of place. Similarly, in a classroom setting, learning math can be a challenge in community because the skill is at the center of instruction. So if group work isn't designed well, students who need time to think will be dismayed by those who finish first. The singular focus on the skill, even when it's warranted, leads people to idolize those perceived to be stronger in that skill, and creates unnecessary hierarchies in any mathematical community. I've heard many people express versions of Simone Weil's despair at "the idea of being excluded from that transcendent kingdom to which only the truly great have access."[8]

Therefore, those who desire community in mathematics must develop the virtue of *hospitality,* which includes *excellence in teaching, excellence in mentoring,* and a *disposition to affirm others.* Hospitable math explorers will bend over backward to reassure newcomers that they are welcome at any stage of development. They will show newcomers the manual of hidden knowledge—which includes the fact that even seasoned veterans don't understand everything in a talk—and mentor them in skills, like how to read math for the big ideas. They will publicly assign competence to others, by recognizing the things they've done well. Those who wield power in mathematical communities must remember that they have great responsibility in setting norms for how to welcome newcomers. Hospitable math explorers will strive to become excellent teachers of mathematics, able to reveal the joy of mathematics even to those who are new to the community. There is so much evidence-based knowledge about good teaching, and we should avail ourselves of it.[9] We can help people gain access to this kingdom through excellent communication.

Those who lead mathematical communities must be practiced at managing group dynamics, with attention to student agency, identity, and power. Skilled teachers know the importance of setting good norms for how people treat one another. They know that group work can be ineffective if one person dominates, and damaging if not every person in the group has a meaningful way to participate in the work. For that reason, math educators stress the importance of designing *group-worthy tasks:* activities in which there are multiple important roles, real collaboration is required, and everyone's contributions are needed for the group to succeed.[10] Effective teachers know how to invite students to share their thinking, and find ways to lower the social risk of participation.[11]

Building community in mathematics involves developing collaborative skills that diminish hierarchies. Successful collaborations are inclusive and benefit from diverse viewpoints. They aren't just a division of labor; rather, the best math collaborations are synergistic, require preparation, and generate deeper understanding when participants push one another toward further growth.

Mathematical communities, like all communities, are prone to implicit bias: unintended, unconscious stereotypes that all of us possess. We make erroneous assumptions about others that affect the power dynamics of the group and limit whose voices are heard. In school settings, we must ask: Who hasn't spoken yet? Whose contributions are often overlooked? At the professional level, we have to recognize how bias leads to decisions that work against community. For instance, when women collaborate with men on research papers, they are less likely to get credit for the collaboration—people think the men did the work. A 2016 study in the similar field of economics shows that while women publish just as much as men, they are twice as likely to be denied tenure, except if they always publish alone, in which case there is no difference.[12]

Therefore, those who desire to build community in mathematics must have a posture of *self-reflection*. We should be mindful of our potential biases so we can mitigate them, and within our communities we must establish good practices and structures that help alleviate bias.[13]

Mathematical communities are deeply plagued by feelings of not belonging. These can take many forms: *I hope no one finds out I don't know very much* (underlying thought: *I don't deserve to be here*); *no one else here is like me* (underlying thought: *therefore no one can truly understand me*); *I'm never going to be good enough* (underlying thought: *to be like the people I idolize*). These feelings can be exacerbated by the hierarchical nature of many communities. Ricardo, as a forty-year-old college student, may feel a mixture of all these. He's from an underrepresented background, in both race and class. He's been out of school for so long that reorienting himself is an adjustment. He feels handicapped by his past. It's no wonder he constantly feels *I'm not supposed to be here.* In fact, many of us feel this way, for one reason or another. I have often felt isolated in mathematical communities, and I still do, even though I'm a well-established mathematician. When I changed research areas midcareer, I spent a semester at a research institute to try to forge connections in the new community. I often felt out of place, because I didn't know much about the new area, and I come from an institution that's quite different from where other research mathematicians come from—because my college prioritizes teaching. No one knew me well, and I was less likely than others to be invited to social gatherings. To be fair, if others had known how I felt, I'm sure they would have reached out more. That's why being welcoming requires active attention.

So, in addition to hospitality, those who value community in mathematics must develop the virtue of *attention to people.* It means looking at others and seeing beyond who they are just mathematically, especially the young, the new, or the forgotten. This is a virtue you must develop even if you are the newcomer. It took some reflection for me to realize, when I felt unnoticed, that there were probably others who felt the same way. Indeed, we were all newcomers at this institute, where people stayed for only a short duration. As the newcomer, you can notice the other newcomers around you, and you can extend a welcome too.

And leaders of any mathematical community should consider the virtue of *vulnerability.* Leaders who can share their journey and its difficulties can help set the tone for others to do so as well. Teachers who ask students to write their "mathographies"—their autobiographical journeys in math experiences—often begin by sharing their own. Leaders who are vulnerable can help others overcome their feelings of being an impostor. Karen Uhlenbeck, a winner of the Abel Prize (a Nobel-like prize in mathematics), acknowledged: "It's hard to be a role model . . . because what you really need to do is show students how imperfect people can be and still succeed."[14]

One of the great joys for me when I talk about human flourishing in mathematics is that I often hear from people about their own deep experiences. The mathematics professor Erin McNicholas recounted a time when she was feeling anxious about an external event and was pulled into a moment of joy with several students and another professor:

> It was hard to imagine how I was going to break out of the swirl of worry, fear, and anger that jumbled my thoughts. And then . . . I happened to pass a student in another professor's Real Analysis class. I had been talking to an advisee about a problem on that week's analysis assignment. She

> had found a hole in her approach to one of the problems that neither she, nor I, could find a way to fill. So I asked this student if he had solved the problem. He had, and in the same way as my advisee, only he had not noticed the hole. So I asked him about it. Within 20 minutes, a group of five analysis students, the other professor, and I had all gathered and were working together to find a solution to this one problem. We would almost have it resolved, when another issue would spring up. Finally, through the contributions of everyone there, we figured it out. There was a rush of excitement, and the student who was writing up our comments on one of the glass boards did a little dance of joy as he hurriedly jotted down the statements that completed the proof. We laughed with him, giddy in our shared victory.

She said it was the sound of the combined laughter that jolted her and made her realize that she had not thought once about her worries during the thirty minutes when they were problem solving together. Mathematics was a refuge from her worldly concerns and a source of joy because of this spontaneous community that formed. In her story you see so many of the things that exemplify a healthy mathematical community. There was a lack of hierarchy. Everyone had been fooled by this problem, and the professors were modeling that it's okay—and even exciting—to be in this posture of not knowing but wanting to find out. They were driven by a common curiosity: even though the students knew that they wouldn't be marked down for not solving a problem that the professor had not solved, they, like their professors, needed to know the truth. They felt a shared hope of solving the problem, and when they did, they were rewarded with joy. Erin reflected on this experience:

> We were all contributors to the solution. I was attempting to contain my pride in my advisee who initially noticed the hole in the argument. While I see her unusual talent for critical introspection, I think her talent is more easily overlooked than the mathematical creativity and intuition exhibited by some of our majors. She has these talents too, but humility and an ability to recognize what you don't know can silence these talents in the group setting.
>
> If we think of a mathematical challenge as a river to cross, some mathematicians' approach is to bound from the shore, jumping from rock to rock, worrying only about the next foot placement. Others pause at the bank, looking for a path across, calculating currents and the likelihood of slipping, looking on Google Maps for a possible bridge somewhere up- or downstream. It is easy to marvel at the bravado and daring of those intrepid rock jumpers, but often it is the diligent planners that come to their rescue when they find themselves stranded midstream.
>
> As a community, as professors and fellow students, I think we overlook the careful, methodical work that ultimately gets us to the other side. I couldn't help but revel in the fact that two professors with PhDs and several senior majors had completely overlooked the same gap in argumentation, and that this student who often does not receive the recognition she deserves spotted it.

This is a picture of a flourishing mathematical community: people who have joined together in a common mission of exploration and play, bouncing ideas off each other, valuing one another's input, getting excited about the directions their ideas are taking them in, and embodying a wide array of mathematical virtues along the way.

FIVE POINTS ON A SPHERE

Given any five points on a sphere, show that some four of them lie on a hemisphere that includes its boundary.

This is a great problem with an elegant solution.[a] So it's a nice final puzzle for this book, because it showcases what makes a beautiful problem: it's easy to state, it has a surprising conclusion, there are multiple ways to explore it, and you can think about it during idle moments of your day. Remember that math explorers are comfortable with struggle, and it's okay to just let a problem marinate in your brain. After a long struggle, when you finally discover a solution, you'll be delighted.

a. Putnam Mathematical Competition problem, 2002.

July 25, 2018

I've actually gone through and studied the whole sections [in Stephen Hawking's *God Created the Integers*] on Euclid's "Elements," Archimedes' "Methods," "Sand Reckoner," "Measurement of a Circle," and "On the Sphere and Cylinder," and I've also gone all the way through Descartes' "Geometry." Since then I've been through Lobachevsky's "Theory of Parallels" and now am almost through a first sweep of Bolyai's "Science of Absolute Space." And honestly, I didn't know Geometry was that rich. I completely understood all that Euclid and Archimedes had to say, more than 90% of Descartes' work, about the same with Lobachevsky, but Bolyai's is somewhat harder (but more detailed) and he is somewhat ascetic and austere with some of his notations and concepts, but I definitely understand the vast majority of what I've gone over so far, and I plan to go over all the parts that weren't crystal clear when I finish. This book "God Created the Integers" is *great.* Do you know of another book like this? That's a compendium of mathematics from ancient up to modern times? That would be a great help. This book here over the last 4 months has really augmented how I see mathematics. . . .

Your phrase "the humanity of doing math" really made me think. Back before I got really serious about mathematics, I used to play a LOT of chess, so I tended to analogize a lot of life to chess. But now that my main focus is mathematics, a lot of mathematical concepts creep into the way that I see a lot of things. . . . I have a friend that I left at the medium security institution I was at, who used to tell me all the time that it takes tenacity to attempt what I'm trying to do. Mathematics as a class in Perseverance?? But one thing I also noticed was in this book [is that] most of them within the same time period knew each other and communed with each other and/or trained each other or were trained by mathematical descendants of one another even throughout time periods—there was a definite *human* structure to their mathematics.

Anything that you write is an inspiration to me and I can't wait to read your book. Yes, it would be OK with me to tell more of my story in your book. If some of the conversations we've had about mathematics help you to illustrate a point, please use them. If I have my way (which I absolutely believe I will), whether it's a couple years from now or 15 or more years from now, I intend to use my story to help someone like me 15 years ago or to help someone else help that person or people.

Because I truly believe if someone like me . . . had stepped into my life when I was 17 and had just gotten my G.E.D. and was enrolled in Atlanta Tech (when I was at that crossroads) and really showed me some of the better things life had to offer (or explained these things), I would've had a way higher chance of doing other than I did. In my opinion (this is a *radical* oversimplification of what I'm trying to say), people don't care enough about other people right now (and haven't for a *while*), from my personal experience and the things we see in the world.

Chris

13
love

I may be able to speak the languages of human beings
and even of angels, but if I have no love, my speech is
no more than a noisy gong or a clanging bell.
Paul the Apostle

We must remember that intelligence is not enough.
Intelligence plus character—that is the goal of true education.
The complete education gives one not only power of concentration,
but worthy objectives upon which to concentrate.
Martin Luther King Jr.

Graduate school in mathematics was crushing my spirit. After toiling for two years to try to solve a particular research

problem, I discovered a fundamental error in a paper that I had based my ideas on. My work was worthless. To salvage something to show for my efforts, I thought that maybe I could publish the counterexample I had found. At least if I did that, I could show the world that my work had been worth some meager thing. But then I discovered that someone had already published the same counterexample twenty years earlier, in an obscure journal I had never heard of.

So much of my identity was wrapped up in getting a PhD that to give it up at that point would have been nearly unthinkable—yet there I was, thinking about quitting math altogether.

Since my youth, I had savored number patterns and liked stretching myself with challenging puzzles. I read popular math books by Martin Gardner for recreation, and I loved being a math explorer. In high school, I remember opening a graduate text in mathematics and not being able to understand any of it, but possessing an intense desire to make sense of it. I dreamed of getting a PhD in mathematics. My parents had that dream for me as well. Education was my parents' marker of success: they were immigrants who had come to the US from China, and they worked odd jobs to make ends meet while pursuing advanced degrees. Needless to say, they were thrilled when I got into Harvard. But my mother, who was battling Lou Gehrig's disease, wept bitterly when I went to Boston, since I would be so far away from our home in Texas. I experienced a severe depression. I found these entries in my journal from my first two months at Harvard:

> I feel lost at the moment. Trapped—because my family wants me here, and I feel like I should be at home, trying to help in some way.
>
> Self-doubt is a rather ubiquitous experience among the first-year students, but I feel like I've got it in some dispro-

> portionate measure. The workload that I've had is tiring, and I wonder why I'm not attacking it with the same enthusiasm as I think I would. I can't seem to pinpoint what the underlying problems are.
>
> I seem to be questioning whether I want to be a mathematician, although I cannot see myself doing anything else.

Those feelings didn't let up, and I struggled for three more years with doubt. In graduate school, the normal thing to do is to apprentice with a professor (your advisor) to write a dissertation (an original, new piece of research) so that you can get your PhD. I worked with one advisor, then another. I could tell that neither thought so highly of me. By that point, I had stopped believing in myself.

Why was I trying so hard to get this PhD? What did it mean to me? Wrestling deeply with that question, I had to ask a more fundamental question, the one that I opened this book with.

Why do mathematics?

Was it for prestige, some external good? Was it to prove that I was better than someone else at mathematics? Was it to compare myself to others? Was it for significance? I had to honestly confess *yes* to all those questions.

For years, I had found my identity in this thing called mathematics. As a result, I was mildly arrogant when I was outperforming everyone else in high school and college, but now, on the other side of the coin, it was depressing to compare myself to everyone and be found wanting. I now knew how Simone Weil felt comparing herself to her brother André. I no longer smelled the aroma that had drawn me to mathematics in the first place; I tasted only its bitter remains when it is prized for only its external goods. I had lost my joy.

Mathematics was a beautiful thing, but I had made it an ultimate thing. My math achievement could have simply marked progress, but it had become a stamp of self-importance. My math training could have supplied me with a modest confidence, but it only sowed doubt when I used it to compare myself to others. I'd been groomed by society to see math as a way of drawing a circle and putting myself in it, believing that math was a showcase for flaunting talent rather than a playground for building virtue. Each time someone confessed their math sins to me—"You're a mathematician? I was so bad at math"—and I enjoyed the attention, we both took one more step down the road to idolatry, in seeing math as reserved for "geniuses." And now in my struggle, this idolatrous view could convey only one verdict: *you're not one of them.*

The math god promised significance, but it only rendered judgment. When I realized that, I knew I should walk away. I didn't need this math PhD to give me dignity.

I began looking around at other things I could be doing. At the time, careers in finance were popular. I interviewed. I would get asked questions about mathematics, and in explaining things to others, I began to remember that math was—actually—delightful and wonderful. In some of these interviews, they ask you to solve math puzzles to see if you know how to think. And I liked puzzles. I remembered how fun it was to play—especially when there was nothing at stake. I conceded that I would miss mathematics if it wasn't part of my life in some way. I smiled when I realized that.

At the same time, I was a resident tutor in mathematics in one of the undergraduate Houses (Harvard's word for dorms). My job—even though I was going to quit mathematics—was to convince others how marvelous math was. The simple rhythm of meeting with and caring for others took my eyes off my own distress. Here were earnest souls, quite capable in math and interested in math, not believing they could flourish because they didn't stack up to others. They were deeply discouraged, just as I was. I would reassure them that they did not need to compare themselves to others to find worth and dignity in themselves or in mathematics, while inside, I would say: *I need to know that for myself.*

Some of the best experiences I had in graduate school were times when I sat over the tutoring table and used the vehicle of mathematics—its wonder and its delight—to care for another human being. I enjoyed getting to know people through mathematical pursuits. You get a similar feeling when you play a sport with someone—you know them in a different way. I felt it an honor to walk with them in their mathematical struggles and to counsel them to see themselves differently. I loved watching people light up when they grasped an idea. It's one of the best feelings in the world.

The love I want to discuss in this final chapter is not the love of math itself. For if you've come this far with me, I hope you've already begun your journey to love and explore mathematics. I'm also not going to talk about using math to analyze the ways that people love, although there are some interesting mathematical models for that.[1]

The love I want us to dwell on is the love one might have for another human being *through and because of* mathematics. Love, the desire, is met by love, the virtue.

Love is the greatest human desire, for it serves all the other human desires—for exploration, meaning, play, beauty, permanence, truth, struggle, power, justice, freedom, community—and love is served by them. To love through and because of mathematics is to build community for the isolated, to seek justice for the oppressed, to help one another grow through struggle, even in mathematical ways. To love is to give the gift of play and exploration, to grow in a desire for truth and beauty, to bestow creative power on another human being by showing them mathematics. To love someone is to set them free, not just in their heart, soul, and strength, but also in their minds.

And love is the source of and end of all virtue, for it sits at the heart of every virtue—even the ones that mathematics builds. To love through and because of mathematics is to build hopefulness, to cultivate creativity, to promote reflection, to foster a thirst for deep knowledge and deep investigation, to encourage in ourselves and one another a disposition toward beauty and all the other virtues we've discussed.

To love and be loved is a supreme mark of human flourishing.

But what kind of love?

There is the false hope of conditional love, the fleeting love that depends on feelings of the moment, a transactional kind of love that has little effect in mathematical spaces because it's the way we've always operated. Society bombards us with this message: value the rich, the strong, the well educated, the powerful. And sadly, it is no different in mathematics classrooms or in the home. The ones who are already able attract attention—they are the ones we see and notice, the ones we admire and believe will do great things. And they succeed in mathematical ways, because we believe in them. But what of the others?

I'm not saying that mathematical expertise shouldn't be valued, nor am I saying that accomplishments shouldn't be recognized. These accomplishments represent the best of what human beings can achieve, and they should be celebrated as a credit to us all. A proof of an outstanding conjecture or an amazing application of mathematics should be announced from the rooftops just like any record-breaking accomplishment in sport. But we should remember that the individuals who make such discoveries stand on the shoulders of others, many of whom go unnamed and unnoticed. The success of these individuals is a testament to their hard work and initiative, but it is also a testament to the investment of the community around them and a product of the blessings in their lives, which were largely out of their control. So achievement in large part belongs to the community, and is a product of who we invest in.

I say this to encourage us to nurture the potential in those who are easily forgotten, including ourselves. Don't the forgotten among us have just as much dignity as the one who is accomplished? Aren't they worthy of our attention? Can't we encourage the one who is at the start of his or her mathematical journey? Shouldn't we embrace the forgotten one who has not experienced the opportunities that others have had for flourishing in mathematics? Wouldn't it benefit us to learn from those persons most different from us, to honor their ideas, to have a posture of humility and not arrogance toward the experiences that they bring?

This is unconditional love. Only this kind of love has the promise of changing the practice of mathematics from a self-indulgent pursuit to a force for human flourishing. Unconditional love recognizes that each person has a fundamental dignity, which does not come from anything they do. Unconditional love reminds us that every person, weak or strong, from every walk of life and every category of uniqueness, is worthy of our time and attention just because they are there, sitting before us. Unconditional love reminds us that to love someone is to really *know* them, to get to know not just their mathematical selves but their whole person.

Too often, those of us who teach math professionally say,

"My job is to teach math," as if teaching math were only about teaching facts and procedures. We forget that "my job is to teach people" whose experiences often interact with mathematics in completely different ways than our own experiences do. And that means education has to take into account the whole person: the joys and sorrows of each person beyond just the mathematics they are learning.

As a mathematical learner, don't let yourself be sucked into an education that champions mathematics as pure logic, cold and heartless, a bunch of rules to follow. Who would want to learn that, or teach that? That is not where the heart of mathematics is. You cannot separate the proper practice of mathematics from what it means to be human.

Because we are not mathematical machines. We live, we breathe, we feel, we bleed. We are embodied human beings. Why should anyone learn mathematics if it doesn't connect deeply to some human desire, something we long for—to play, seek truth, pursue beauty, find meaning, fight for justice? You, as a mathematical explorer who is learning to embrace your identity in math, can be part of the movement to read people differently.

Believe that you and every person in your life can flourish in mathematics.

This is an act of love.

For you love another human being when you honor them as a dignified mathematical thinker and believe in their potential to realize their mathematical brilliance. You love yourself when you grasp a vision of the fullness of your own capacity and refuse to let anyone deter you from claiming a mathematical heritage that is humanly and rightfully yours. You love all people when you stop speaking about talent as a thing you either have or don't, and start speaking of the virtues that each person can build—through the hope and the joy of courageous effort and hard work. To love is to believe that everyone can flourish in mathematics.

And this is a challenge for all of us. Because I've fallen short of this ideal. I've written off students, sometimes unintentionally through unconscious bias, sometimes intentionally because I don't have the imagination that I need for the sacred responsibility of teaching.

You may have a Ricardo in your life, who lacks a mentor to guide him in math and science. *You* can be his constant encourager. You may have a Simone in your life, who's always comparing herself to the Andrés around her. *You* can help her establish her own identity in math. You may have a Christopher in your life, fallen into drugs and in with the wrong crowd, who never seemed interested in math, maybe even seems lazy. If you knew what he was going through, maybe you'd read him differently.

> *Believe that you and every person in your life can flourish in mathematics.*

Six years after Christopher wrote me that first letter from prison, he's helping other inmates learn math to get their GEDs. He's using the meager income he makes to buy math books, and he's studying topology and advanced analysis now. He says:

> I'm studying anywhere between 3 and 5 hours M–F and 2+ hours Sat. and Sun. depending on how I'm feeling. It's harder to study and read here because you're not in a "traditional" cell with a cell door that you can be enclosed in, but everything is open and we live in an open-top 8×10-foot "cube" with walls that extend 6 feet off the ground and a 3-foot gap in the 8-foot side to act as your "door." Also I'm in a room without a "desk," so I have to take two chairs to do studying in. But hopefully that will change

soon, because I'm in the works of trying to move to a cube with a desk.

But I can't complain too much. I grab my earplugs, my two chairs, and go to work.

No one would call him lazy or disinterested now. Would I have had the imagination to see a future for him fifteen years ago? Where would he be if he had been shown then, in Simone Weil's words, "that transcendent kingdom"?[2]

Believe that every human being can discover an affection for mathematics.

Invest in someone who you know is facing challenges, and become their long-term advocate, in mathematics and in life. You can be a mentor, an encourager, a cheerleader for someone in their mathematical struggles—you don't have to have any advanced background to do this. Consider doing this for those most easily forgotten. Be the one who says, "I see you, and I think you can flourish in mathematics." Be the one who searches out opportunities for them. Be the one who guides them to virtue. Be the one who sees their hardships and asks, "Is everything okay? What are you going through?"

Believe that you and every person in your life can flourish in mathematics.

Every being cries out silently to be read differently. Every being cries out silently to be loved. Christopher, in prison, wasn't looking for only mathematical advice. He was looking for connection, for someone to reach out to him in his mathematical space and say: "I see you, and I share the same transcendent passion for math that you do, and you belong here, with me."

When I was in the depths of despair in graduate school, struggling against professors who thought I would never succeed, one professor reached out to me and became my advocate. After I told him that I might leave, he said: *I would rather you work with me than quit.* It was an act of grace—undeserved kindness—indeed love, that he reached out to me.[3] Through my soul-searching, I had already freed myself from the burden of needing the PhD to give me dignity. Now I was being offered a chance to come back to mathematics, this time for the joy of mathematics itself.

For in the exercise of the mind and the openness of the heart, two people can truly see each other—without judgment or shame—when they behold the same beautiful truths, or dream in hope under the same starlight. In wanting the best for each other, they will each go to great lengths to help the other see more. Each of us is limited by circumstance, but not by imagination. None of us wants to be written off or forgotten. All of us want to be read differently, in all our truth. The starry spheres we marvel at, the exquisite patterns that call to us, the celestial symmetries we wish to explore—these are treasures that belong to all of us, unlocked for us by people who love us and believe in our ability to cherish these gifts.

So I ask you:

Whom will you love, whom will you read differently?

I'll end with a few reflections.

The first is by Simone Weil. After wrestling with her insecurity in mathematics, she saw that there was a path to virtue through her struggle and that it could help people. She wrote:

> The love of our neighbor in all its fullness simply means being able to say to him: "What are you going through?" It is a recognition that the sufferer exists, not only as a unit in a collection, or a specimen from the social category labeled "unfortunate," but as a man, exactly like us, who

> was one day stamped with a special mark by affliction. For this reason it is enough, but it is indispensable, to know how to look at him in a certain way.
>
> This way of looking is first of all attentive. The soul empties itself of all its own contents in order to receive into itself the being it is looking at, just as he is, in all his truth.
>
> Only he who is capable of attention can do this.
>
> So it comes about that, paradoxical as it may seem, a Latin prose or a geometry problem, even though they are done wrong, may be of great service one day, provided we devote the right kind of effort to them. Should the occasion arise, they can one day make us better able to give someone in affliction exactly the help required to save him, at the supreme moment of his need.[4]

She had found a path through struggle to virtue. She understood that mathematics is for human flourishing.

This is from Ricardo Gutierrez, the forty-year-old audio engineer from New York City who returned to school to pursue math and computer science:

> I'm in classes with 20-year olds. I'm having the time of my life. . . . The learning has unlocked much I didn't know existed in myself.
>
> Since I've been back I've struggled with math. Calculus has really beat me up. After a 20-year break from it I'm finding it harder to relearn, finding it impossible to imagine I was ever really good at this. But even in the pain and failure of trying to reshape my brain to comprehend, I feel more alive than I ever have before.

Ricardo isn't letting his struggle undermine his goal of pursuing mathematics. He understands: mathematics is for human flourishing.

And this is from Max Triba, a data scientist, who wrote to me after reading the "Flourishing" speech I gave as the president of the Mathematical Association of America:

> I just finished reading your wonderful article "Mathematics for Human Flourishing" and felt compelled to reach out with a small personal story. When I was in second grade, I struggled with subtraction and asked my teacher for help. She snapped, told me some mean-spirited equivalent of "You need to go figure this out because it isn't hard," and I returned to my desk feeling like the biggest idiot. I barely ever asked for math help after that and struggled for mediocre grades until college.
>
> In college I fell in love with an aerospace engineering major, and her deep understanding of mathematics was almost intimidating. And at the same time I discovered a passion for economics and through that, math's ability to elegantly explain complex phenomena. I only have an undergraduate degree, but I've managed to work in applied math ever since graduating and today do time series analysis in healthcare. If only I could tell 8-year-old me of this trajectory.
>
> Discovering the beautiful intersection of mathematics and humanities will always have a very special place in my heart, and I love to share it with others. The path I've been on has shaped my perspective that anyone, regardless of gender, ability, race, or otherwise, can be a part of this wonderful thing.

Max, who was initially turned away from math but was brought back through love, now understands: mathematics is for human flourishing.

My joyful hope in writing this book is that I've given you encouragement in your own journey in mathematics. I hope you now never feel like saying, "I'm not a math person," because you are a human person, and you can see how doing math is tightly bound to being human. I hope I've equipped you to speak about mathematics with others as a vitally human endeavor, grounded in basic desires that we all share, and elevated by virtues to which we can all aspire. If we embrace this vision, we can indeed love each other better and help one another grow, through and because of mathematics.

I bid you shalom and salaam, grace and peace, in all your explorations. May you—and all those you are called to love—flourish.

May 31, 2017

Over the last two years, I've been working as a G.E.D. mathematics tutor at this institution.

The education department here operates highly inefficiently, but I've still managed to help 12 of my fellow inmates to achieve their G.E.D. credentials. I have a young guy who is going to be released in a couple of years now, and he says he wants to go back to school to learn engineering when he leaves. So over the next two years, I'm endeavoring to help him learn Algebra II, College Algebra, Geometry, Trigonometry, Calculus I, and Calculus II.

I recently read an article in a newspaper about a woman who was twenty-six who decided to go back to school to become an engineer even though she wasn't particularly good with mathematics. Sometimes your spirit isn't in the best of places in a circumstance like mine. But I am now totally encouraged to redouble my efforts to pursue my goal of becoming a mathematician so as to be able to teach and study mathematics one day after I'm released. . . .

Studying mathematics has given me the opportunity to progress into a person who is better than who I was, toward a person who will be even better in the future, and down a path that I can and fully intend to happily and faithfully follow to its end.

Chris

epilogue

FRANCIS: Chris, it was generous of you to allow your story to be told in this book. I've appreciated our correspondence over these many years, and I love the fact that others will be inspired by you the same way I have been. May we chat a little more about your experiences for the benefit of our readers?

CHRIS: Sure. It was generous of you to share my story. I've learned a ton from our correspondence, and if it inspires others, that would be great.

FRANCIS: Our readers have seen your journey through mathematics, first through elementary texts, and later through

more advanced books in math. Now you are reading journals that professional mathematicians read, even when you don't know all the terms yet. You have a persistence to read advanced papers that I don't think I had when I was at your stage of learning mathematics.

CHRIS: For me math is like a doorway to be able to create things, maybe like Minecraft or something like that. I like abstract things and it seems like a metaphor for so many things. It has power, it has breadth, and it seems to link so many things so completely. What do I see in it? Well, let me put an example like this: Study a text in logic, study it and really understand it, and then go have a debate with a person you usually converse with (if you have a good debate partner). If they're making an argument that's completely illogical you can literally "see" why their logic is completely broken, and you can explain it to them. (Doesn't necessarily mean you'll persuade them or win the argument, though.) And that's just dealing with logic.

FRANCIS: You know I like to ask this question: what have you been learning about the process of doing or creating math?

CHRIS: That there is a lot of different ways to do it and you should choose your best way. Sometimes it's an off-the-wall idea, or an out-of-the-box idea or a counter-logical idea, but they all can work. It definitely takes more than mere concentration. Mathematics takes creativity.

FRANCIS: Yes, and indeed, exploration stimulates creativity. Readers have now seen my argument for mathematics as a force for human flourishing, universal in the way it can address basic human desires, and beneficial in the ways that a proper practice of mathematics builds many kinds of virtues. I'm wondering in what ways you've been encouraged or challenged by this message, and what virtues you feel your pursuit of mathematics has built in you. You played a critical role in reading through drafts as I was writing (thank you), so I know you've thought about this stuff deeply.

CHRIS: A lot of the stuff we talk about encourages me and challenges me (I believe both of these things are not too different). Most all of the book encouraged me and challenged me, but there was one line that particularly illuminated some things for me. "Creative power is *humble,* and it puts others first. It seeks to unleash creativity in others." When I first really took hold of my agency, I really actually recognized that I'd done a lot of wrong. And now that I'd like to do right (a lot if I can), I've really started paying attention to the things that I did and how they affected the people and circumstances around me. When I started to teach people, I couldn't put into words exactly what I was trying to do but "it seeks to unleash creativity in others"—that put more of a light on it for me.

Mathematics has helped me better fortify my patience: sometimes when I'm dealing with some frustrating situation, I remind myself I must have infinite patience. I've experienced hope that a solution will come to a problem also. I've had the experience of: right now, I just can't see the solution, but if I walk away and come back, I might see the solution, or the next day, or next day, but if I keep it on my mind, I'm going to get it. Nine times out of ten this has worked so far. And also, I understand the community: none of us can be great if we're not willing to teach each other.

FRANCIS: So true. Of course, I try to make clear in the book that I don't think of math as a panacea to address every ill. It won't solve every human problem, and it's not a spiritual answer to the ultimate purpose of humankind, but it does contribute in important ways to a life well lived, and your experience exemplifies that. What are some of the other things you pursue that contribute to your own flourishing?

CHRIS: Well, I've learned a lot playing chess. And working

out also puts an endurance on me—especially running, where almost every half-mile I want to stop. But if I keep resisting that urge, before I know it, I've run four, five, or sometimes ten miles. Actually, doing a lot of listening to a lot of different people has taught me a lot also (actual listening and deep engagement with what they're saying). It also allows me to always try to remember to change perspectives (like you always say) and it keeps me open-minded.

FRANCIS: I know people are going to take a lot of inspiration from your letters. They were all written before we decided to include them in the book. What was it like for you to reread them?

CHRIS: Well, it allows me to change perspective and think about and critique what I said. I like to go back and look at what I've written (poems, letters, etc.)—and what people have written to me for that matter—and see how the relationship has changed, how the person has changed and how I've changed. I can see from our first letters and all our other correspondence how my knowledge has gotten a little deeper and broader.

FRANCIS: I can see that too. I'm excited for you. And you are also now teaching other people math in prison. What are some of the things you say to others to help them see math in a different way?

CHRIS: I teach guys that have been convicted of crimes—a lot of us have sold drugs in the past—so explaining things in a buyer/seller terminology tends to break through well a lot. I recently was explaining slopes of lines, constant rates of change, and linear functions, and I always tend to say things like "x is independent, y is dependent, x is time, y is money . . . if you sold 7 shirts an hour, how many did you sell in 3 hours . . . 21. What about 4 or 5 hours? That's a constant rate of change." Stuff like that, any time I can make it "real life" for them, that's what I try to do.

FRANCIS: Readers may have only inferred it from the letters, but you've had a really hard year, and suffered some injustices at the last prison facility you were in. I'm really glad things seem to be better now. What's been the most challenging thing about life in prison?

CHRIS: Well for me—I'm a willful person—the loss of your self-dominion is an obvious one. In most cases, you don't even have control over your own time (as far as how you spend your time in a day), so that's hard. You're sentenced to a time of confinement—not to also be treated like you're less than human—some (nowhere near all) staff see you like that. As far as what it does to your spirit, my spirit hasn't been in too much of a dismal condition for a while, but I've been there. The monotony, the forced listlessness, it's a "provisional existence" (I forgot where I read that).

Put it like this: if you're a person who wants to do something with your life, it definitely can get to you—it's like a coercion into a purposeless, meaningless existence. Math has helped me tremendously by giving me goals to focus on, and even broadened out to other goals in teaching people mathematics and getting people more interested and engaged in education in general.

FRANCIS: In the book I spend some time discussing race. It's not an easy topic to discuss, because people have so many different experiences related to it, some of them painful. But my hope was to encourage self-reflection about what assumptions we make about people with regard to mathematics. What obstacles have you encountered as an African American, either in education or in life more generally?

CHRIS: I can't say that I had any necessarily bad experiences in school when I was young (unless I did something to cause it). I actually went to some decent schools when I wasn't

After five years of correspondence from opposite sides of the country, Christopher Jackson and I met in person for the first time in November 2018. The background is a mural painted on a prison wall, the only place where prison administration would allow a photo.

in alternative school, and had good teachers of all races. I've been in jail my whole adult life, so most of my experiences are jail experiences, and when I was out in the world as a teenager I lived a life that—if you haven't lived like that, or closely known someone that has, it'd be hard to understand. Heck, I've gotten to the point where even I don't understand it sometimes. But since I've been in these low-security institutions, I've experienced racism from other non-black inmates, and also from older black inmates and staff. When you're young (well, I still look fairly young) and a black male, everybody expects you to not know anything. It's hilarious to me. I should probably be offended, but look where I'm at, and I also understand that most people only understand what they see or what they think they see at least, so I get a laugh out of it. It doesn't bother me too much.

FRANCIS: I was glad we got to meet in person for the first time a few months ago, after so many years of correspondence. I know I was a little nervous about what to expect, but really looking forward to it. When you've only known someone through their writing, it can be kind of strange to meet in person. But I think we broke the ice quite well by playing two games of chess. You roundly defeated me both times! What was our first meeting like for you?

CHRIS: You weren't too nervous, at least I couldn't tell, and you told me you haven't played chess in a long time either, so that counts for something too. I really enjoyed talking to you. I know I speak with a lot of fervor. Guys tell me all the time: "No, Chris, you're not aggressive, but you definitely speak like you really believe what you're saying." When we did break the ice, and we were both talking animatedly, I learned a lot (all squares are congruent to 1 or 0 mod 4). You walk it like you talk it, humble, motivating, and generous. I like that and I'm striving to emulate it.

FRANCIS: I can say the same for you—you're earnest, thoughtful, humble, and generous too. What advice would you give a younger version of yourself?

CHRIS: Younger Chris, I'd have to talk to him kind of rough to start with, but once I got him to listen to me, I believe he would have. I always had a tendency to listen to the guys a little older than I was, especially if I respected them and it seemed like they had a little sense. It would have had to be a continuous talking to him because he was so hard-headed. But my main message to him would be "It's far more important to understand everything going on around you than it is to be cool or to fit in with your surroundings, and the world is way wider than you think it is: there are an almost uncountable multitude of ways to achieve what it is that you think you want to achieve at way less risk to your life and future. And nobody's life or future is worth any amount of money or other thing that people like us tend to chase that are in reality insignificant." Younger me was extremely headstrong, which developed into strong will in me today, but I believe a person like me now could have reasoned with him then.

FRANCIS: What are your hopes and fears going forward?

CHRIS: I've done a little good; I hope to be able to do more as my life goes on. I hope to be able to reach young people like I was, so they don't have to go through the things I've gone through, do the things that I've done, or take their families and loved ones through what I've taken my family and loved ones through. I hope when I get free, to be able to get more people excited about education, and I hope I'm not delusional, but to get more people to understand and use the power that they have. My fears? I could say I fear that I won't succeed, but that wouldn't count because we all have that fear, so it's not particular to me. But I can share a fear I've had before. I've feared before that I'd be here so long that total cynicism would overtake me, and I'd lose the energy to be able to go forth and do the things that I want to do when I leave. I don't think that's much of a possibility anymore. But I don't really have any fears. I feel that if I can make it through this, I can make it through anything else that is put in front of me in the future.

FRANCIS: Chris, thanks for opening up to our readers about yourself and your story. You are flourishing even in the most difficult circumstances.

CHRIS: Thanks for helping me to tell my story. I've been fortunate: the people I've been around, and who have stayed in my orbit while I've been striving, are helping me to flourish. Thank you.

Chris was first incarcerated at the age of nineteen. He has served thirteen years of a thirty-two-year sentence. There is no parole in the federal system. With good conduct time allowance, the earliest he could be out is 2033. His sentence was for two convictions, either of which alone would have given him seven years of time, but overly harsh sentencing laws required the second offense to add twenty-five years to his first seven years. The First Step Act, passed by the US Congress in 2018, reduced sentences for offenses like the ones he was convicted of, but not retroactively. If it had, Chris would be free now.

For Chris's contributions to this book, he is earning a portion of the royalties. I'm mindful that society often uses the labor of the marginalized without compensation.

Mathematics has helped Chris flourish and live a more fully human life than prison might otherwise allow, and he's helping others to flourish too. There are many Christophers, both inside and outside prison. On my website (francissu.com) I maintain a list of organizations and resources that can assist those who could flourish—and help others flourish—with your support.

There may be a Christopher or a Simone already in your life. You can mutually encourage each other.

acknowledgments

Let me start by thanking you, the reader, for taking the time to sit with this book and these ideas. Simone Weil said, "Attention is the rarest and purest form of generosity." I have found the writing process demanding, but in a good sort of way, as it has forced me to think deeply about things I care about and express them with utmost care. I'm appreciative that you would let me into your own space as you reflect on your experiences. In mathematics, may we never lose sight of matters that speak to who we are as human beings.

I am indebted to Joe Calamia and Yale University Press for believing in a project that isn't a traditional math book. Joe has been such a pleasure to work with—thoughtful, patient, and wise—and gave critical advice at key points that improved what I was trying to say. I'm so grateful to the copy editor, Juliana Froggatt, who made my words sound better than they deserve, and to Margaret Otzel and the others at the Press who did an amazing job producing the book and making it visually appealing. Mad props go to my friend Carl Olsen, who drew the gorgeous chapter-opening illustrations on a tight timeline and accomplished my goal of making a book about math feel human.

I'm grateful to the Mathematical Association of America (MAA) for permission to use my writing as its president as a starting point. Portions of this book are adapted from my speech "Mathematics for Human Flourishing," which was published in the *American Mathematical Monthly* 124 (2017): 483–93, available also at https://mathyawp.wordpress.com/2017/01/08/mathematics-for-human-flourishing, and from a regular presidential column I wrote for *MAA FOCUS.* I cherish the people in the MAA community with whom I've served, who work tirelessly to improve the teaching of mathematics at the college level and value the things I've written about in this book.

I have been shaped by the care of many good people. Many were generous with their attention when the book was in development. Family and friends, most notably my sister Debbie, were of huge support during a busy year that included writing a book and planning a wedding (mine!). Several who encouraged me through times of professional doubt appear in spirit in these pages—including Jennifer Wiseman, Soren Oberg, John Fuller, Rob DeWitte, and Mark Taylor. In more recent years I've been grateful for the friendship of Tom Soong, Zac Marshall, and Phil Cha. I wouldn't be a mathematician today without the support of Michael Starbird, my undergraduate mentor, and Persi Diaconis, my PhD advisor, who kept me from quitting mathematics altogether.

Supportive colleagues at Harvey Mudd College have helped broaden my commitment to becoming a better mathematics teacher. And they are just plain fun to work with! Numerous people read an initial draft of this book and made substantive critical comments that greatly clarified what I was trying to say, including Yvonne Lai, Darryl Yong, Ben Braun, Elizabeth Kelley, Michael Barany, David Williamson, John Cook, Dave Henreckson, Kim Jongerius, Art Benjamin, Tori Noquez, and my freshman writing class. I received especially extensive feedback from Robin Wilson, Russ Howell, Pat Devlin, Ron Taylor, Josh Wilkerson, and anonymous reviewers. It's almost a different book after all their comments. I'm also grateful for helpful conversations with Adriana Salerno, Judy Grabiner, Jon Jacobsen, Rachel Levy, and Michael Orrison. Matt DeLong deserves thanks for help with the MAA speech on which this book was based. Many of the book's ideas popped out in conversation over regular lunches with David Vosburg, who has been such a loyal friend and has provided wise counsel over many years.

Responsibility for the views expressed in this book rests with me.

I can't really express how blessed I feel to know Christopher Jackson, and I consider it a privilege to have a friend like him, shining a light on things I thought I already knew well. I'm grateful for the earnestness with which he read drafts of my book and discussed his ideas with me. Our society needs to do better in opening doors, making paths of redemption, for people in his situation. Excessive punishment and mass incarceration must end. In some small way, I hope this book helps us to see that more clearly.

My loving wife, Natalie, has been a constant source of support and guidance, my partner in this project in so many ways, pushing me not only to be a better writer but to be a better person. I wish you all could see her merciful heart and how she cares for those who have been forgotten by society. Even in our first year of marriage, she considered this book her project also and was supportive every step of the way. She is a true picture of love, friendship, and partnership. We are both thankful to have a faith community that supports us. And as a follower of Jesus, I am grateful to the one who defends the dignity of all human beings and sustains my own experience of human flourishing.

FRANCIS SU
January 2019
Twitter: @mathyawp
Website: francissu.com

desires & virtues

Here is a list of all the virtues mentioned in this book, which are built through the pursuit of mathematics when it is grounded in basic human desires. The desires are the chapter titles, and the virtues I chose to discuss are listed under each.

EXPLORATION

imagination
creativity
expectation of enchantment

MEANING

story building
thinking abstractly
persistence
contemplation

PLAY

hopefulness
curiosity
concentration
confidence in struggle
patience
perseverance
ability to change perspectives
openness of spirit

BEAUTY

reflection
joyful gratitude
transcendent awe
habits of generalization
disposition toward beauty

PERMANENCE

trust in reason

TRUTH

thirst for deep knowledge
thirst for deep investigation
thinking for oneself
thinking rigorously
circumspection
intellectual humility
admitting error
confidence in truth

STRUGGLE

endurance
unflappable character
competence to solve new problems
self-confidence
mastery

POWER

skill in interpretation, definition, quantification, abstraction, visualization, imagination, creation, strategization, modeling, multiple representations, generalization, and structure identification
humble character
sacrificial character
encouraging character

heart of service
resolve to unleash creativity in others
resolve to elevate human dignity

JUSTICE

empathy for the marginalized
concern for the oppressed
willingness to challenge the status quo

FREEDOM

resourcefulness
fearlessness in asking questions
independent thinking
seeing setbacks as springboards
confidence in knowledge
inventiveness
joyfulness

COMMUNITY

hospitality
excellence in teaching
excellence in mentoring
disposition to affirm others
self-reflection
attention to people
vulnerability

LOVE

love, the source of and end of all other virtues

for reflection

questions for further discussion

Our views and our practices of mathematics cannot change without reflection and action. I am providing some questions below as starting points for further discussion. On my webpage (francissu.com), I am maintaining other resources associated with the book that may be helpful to teachers, including reading lists with live links to references.

FLOURISHING

These first three questions are ones you may wish to think about before reading the book, and then return to after you finish it, to see how your answers compare.

1. What is mathematics? How would you describe it to a friend, in a sentence or two? What do you feel is the purpose of learning mathematics, for yourself or others?
2. What connections do you see between doing mathematics and being human?
3. Describe any virtues you have acquired as a result of doing mathematics.

EXPLORATION

1. Think of a time when you were captivated by exploring something (e.g., a location, an idea, a game). What analogies can you draw between doing math and doing this exploration?
2. Consider this statement: "The wayfinders were mathematical ex-

plorers of their society, using attentive study, logical reasoning, and spatial intuition to solve the problems they encountered in their cultural moment." Choose any cultural practice and reflect on ways that mathematical thinking might present itself in that practice.

3. If you teach math to others, what are some ways that you can train your students to expect enchantment?

MEANING

1. "Mathematical ideas are metaphors." Reflect on one mathematical idea that you've now seen in multiple situations, and how the meaning of that idea was enhanced in each encounter.
2. How does abstraction enrich the meaning of an idea? Describe one example from your own experience.
3. "Mathematics is the art of engaging the meaning of patterns." Consider this statement in light of a scientific discovery in which mathematics played a part.

PLAY

1. Think of an activity you associate with play. Make a list of all the things you enjoy about its playful aspects. Does your list have analogies in mathematical activities?
2. Some people seem to have patience and hopefulness when trying to solve a problem, and they will persist for a long time in thinking about it. Others seem to give up quickly. How does math play build hopefulness and patience? Compare this to the discipline of learning a sport.
3. Math play "asks you to change perspective, to look at a problem from different viewpoints." In what ways is this virtue useful in life?

BEAUTY

1. Describe any experience you've had with sensory, wondrous, insightful, or transcendent mathematical beauty. How did that experience make you feel?
2. Think about all your educational experiences—for instance, classes

you've taken in different subjects. Which ones implicitly acknowledged the human desire for beauty?

3. Where is mathematical beauty found in the world?

PERMANENCE

1. What mathematical laws, truths, or ideas do you rely on in your daily life?
2. How is mathematics a refuge? For whom is it a refuge?
3. Many things in the universe change over time (and let's not forget—the subject of calculus was developed to study such change). Do you find it surprising that the laws of mathematics do not change over time?

TRUTH

1. Reflect on a time when shallow knowledge (in any subject) has led you astray. How did that make you feel? How is deep knowledge an antidote?
2. Sometimes parties on two sides of an argument have different perspectives on the same event. Both views may be true, but each may be just part of the picture. How is knowing the whole truth a better place to be? Similarly, in mathematics, what does knowing the whole truth look like?
3. How can mathematical thinking equip you to converse with and respect people who hold different views?

STRUGGLE

1. Describe an activity you enjoy, and make a list of all the internal and external goods you can think of that are associated with that activity. Now think of an activity you don't enjoy, and make a similar list. What do you notice about these lists?
2. What internal goods does mathematics offer? Discuss how these goods multiply when you share them with others.
3. If you teach mathematics, how can you incentivize students to value the process of struggle and not just the outcome?

POWER

1. Think of a recent challenging math problem you explored. Which of the powers of mathematics did you develop or use in that exploration (interpretation, definition, quantification, abstraction, visualization, imagination, creation, strategization, modeling, multiple representations, generalization, structure identification)?
2. Discuss creative power and coercive power that you have witnessed in mathematical settings.
3. If you teach, how do you affirm your students' dignity as creative human beings in the way that they do mathematics?

JUSTICE

1. If people have realized that the way we teach math needs to change, why hasn't it changed yet? Who benefits from keeping it the same way it has always been?
2. All of us unwittingly harbor bias, so how can we mitigate bias in mathematical spaces? Who is harmed by bias in mathematical spaces, and why?
3. What inequities do you notice in mathematical spaces? Who is harmed by those inequities? Think deeper than the obvious answers.

FREEDOM

1. Describe settings in which you've experienced any of these freedoms: the freedom of knowledge, the freedom to explore, the freedom of understanding, or the freedom to imagine.
2. Who do you think may not be feeling welcome in mathematical spaces? In what ways can you extend the freedom of welcome in mathematics to those around you? Think of concrete actions you can take.
3. What things have you experienced in a math classroom that feel like freedom? What things feel like domination?

COMMUNITY

1. Why should hospitality, or excellence in teaching and mentoring, be central to doing mathematics well?
2. How can you build a community in the classroom or the home in which participants push one another to grow while not being overly focused on achievement?
3. What actions can you take to address feelings of not belonging in math communities?

LOVE

1. In what harmful ways do we use mathematics as "a showcase for flaunting talent rather than a playground for building virtue"?
2. How can you honor each person you meet as a dignified mathematical thinker?
3. Who are the forgotten among you, mathematically speaking? Whom will you love, whom will you read differently?

hints & solutions to puzzles

Before looking at any of these hints or solutions, you should first play with the puzzles! Work through examples to get a feeling for what is going on. Take as much time as you want to dwell on the problems—there is no hurry. The struggle itself is valuable.

HINTS

Dividing Brownies. Try special cases. If the removed rectangle is very tiny, how should you orient the cut?

Toggling Light Switches. Try several examples. Look at particular bulbs, and ask which multiples will toggle them.

"Divides" Sudoku. Since within each three-by-three block all pairs of adjacent cells that are related by division are marked with the ⊂ symbol, you can identify the locations of most of the 1s. After that, start looking for chains: for example, if you see A ⊂ B ⊂ C and you know that none of A, B, or C is a 1, then the only possibility is 2 ⊂ 4 ⊂ 8. Also look for cells that divide more than one neighbor or are divisible by more than one neighbor. Notice that 5 and 7 do not divide any other number from 1 through 9 and are not divisible by any other number from 2 through 9.

Red-Black Card Trick. If the black cards from the second pile were replaced by the red cards from the first pile, would the size of the second pile change?

Water and Wine. How is this puzzle like the Red-Black Card Trick?

The Game of Cycles. Explore the game and make some conjectures. It may help to notice: if a triangular cell has exactly one edge marked with an arrow, and you mark a second edge with an arrow pointing in the same direction as the first arrow (so as to partially form a cycle), then the other player can win on the very next move by completing the cycle.

A Geometric Puzzle. Look at the area of the regions where two rectangles overlap. What can you say about this area? Can you cut this region into pieces whose areas are easy to figure out?

Ants on a Log. Suppose you had only two ants instead of one hundred. What can you say about the configuration of ants before and after a collision?

Chessboard Problems. Each domino covers one black square and one white square. If one black square and one white square are removed from the board, must it always be possible to cover it with dominoes? Where does a knight sitting on a white square go after one move? What color squares does a Tetris piece cover? How can you color the little squares in an 8×8×8 cube so that 1×1×3 blocks cover the same number of each color type?

Matsumoto Sliding Block Puzzle. You can construct the sliding block puzzle out of cardboard or paper. How will you get the large square piece past the horizontal rectangular piece?

Shoelace Clock. The answer to part 1 is less than 7.5 minutes. The answer to part 2 is surprising.

Vickrey Auction. Try to show that a bidder who bids her true value *V* will never do worse and will sometimes do better than bidding some other value *B*.

Pentomino Sudoku. Start by looking for double 2s or double 4s in the rows and columns. What does this say, for instance, about the pentomino in the bottom left corner? Counting the number of particular digits required in neighboring rows or columns can be a useful tool.

Power Indices. The set of all orderings of three groups is *ABC, ACB, BAC, BCA, CAB, CBA*. In which of these is group *C* pivotal?

Unknown Polynomial. You need a lot fewer questions that you might first suspect. The fact that the coefficients are nonnegative is important here. Can you bound the size of the largest coefficient?

Five Points on a Sphere. Remember, your goal is to show that no matter what five points on a sphere you pick, there's a hemisphere that contains four of them. You might be tempted to guess the "worst" configuration and show that the claim holds for it, but that is not sufficient to show that the claim holds for *every* configuration. Also, for any pair of points, is there an easy way to ensure that the pair lies inside a hemisphere?

SOLUTIONS

Dividing Brownies. If you cut along a straight line that runs through the center of the large rectangular pan and the center of the rectangular hole, the cake pieces on both sides will have the same area, since each will be half the size of the pan minus half the size of the hole.

Toggling Light Switches. The bulbs that are on at the end of all the toggling are those numbered 1, 4, 9, 16, 25, 36, 49, 64, 81, and 100. These are the bulbs corresponding to perfect squares. To see why, notice that bulb N gets toggled for each switch whose number is a factor of N (a number that perfectly divides N). Perfect squares are the only integers with an odd number of factors. You can see this because most factors come in pairs: if J is a factor of N, then N/J is also a factor of N, and the only instance where J and N/J are not distinct is when $J = N/J$. In that case $J^2 = N$, so N is a perfect square.

"Divides" Sudoku. See below.

3	1	7	2	5	8	4	9	6
9	5	6	4	3	1	8	7	2
8	4	2	7	6	9	1	3	5
5	7	8	1	9	3	6	2	4
6	3	4	8	2	7	9	5	1
1	2	9	6	4	5	7	8	3
2	9	1	5	7	6	3	4	8
4	8	3	9	1	2	5	6	7
7	6	5	3	8	4	2	1	9

Red-Black Card Trick. The reason this trick works can be seen in a couple of ways. Let H be half the number of cards in the deck. If the number of red cards in the first and second piles are respectively R and S, and the number of black cards in the first and second piles are respectively A and B, then we know that $R + S = H$ (since the total number of red cards is H) and $S + B = H$ (since the total number of cards in the second pile is H). Then both R and B equal $H - S$, so $R = B$. Another way to see it is this: if you move the R red cards to the second pile and remove the B black cards from the second pile, the second pile now has all the red cards and is the same size (H) as before. So R must equal B.

Water and Wine. Let H be the total volume of water. If, after this process, the amount of water in the wine glass is R and the amount of wine in the water glass is B, then replacing B by R would leave the volume H unchanged. So R and B must be equal.

The Game of Cycles. The second player has a winning strategy. After the first player begins by marking an edge with an arrow, the second player should respond by marking the only edge in the diagram that does not touch the first edge (the direction of the arrow doesn't matter). After that, as long as the second player doesn't complete the second edge of a potential cycle cell, she will win. There are many other good questions to explore. For other starting diagrams, who has a winning strategy? Are there starting diagrams for which it is possible to mark every edge with an arrow and have no cycle cells?

A Geometric Puzzle. Each pair of rectangles overlaps in a quadrilateral Q. Cut Q along a line from P, the point where the three rectangle borders meet, to the point M, the other point where the two rectangle borders meet. This splits the overlapping region into two triangles that have the same area (by symmetry) and are each corners of a rectangle. It can now be seen that those corner triangles are 1/8 of the total area (4) of a rectangle. So the area of an overlapping region is 1, and there are three overlapping regions. So the total area covered by all three rectangles is 3 times 4 minus 3, which is 9.

Ants on a Log. While ants bouncing off each other seem difficult to keep track of, one key idea makes it quite simple: two ants bouncing off each other is *equivalent* to two ants that pass through each other, in the sense that the positions of the ants in each case are identical. So you might as well think of all the ants as acting with independent motions. From this viewpoint, the longest that you would need to wait to ensure that all the ants were off is the amount of time needed for a single ant to traverse the length of the log, which is 1 minute.

Chessboard Problems. Chapter 6 contains a proof that when two squares of the same color are removed from a chessboard, what remains cannot be tiled by dominoes. If two squares of opposite colors are removed, the remainder *can* be tiled by dominoes—to see this, find a continuous path, moving from square to adjacent square, that visits every square on an 8×8 board. Removing one black and one white square breaks this path into two paths, each of which has an even number of squares, and each path can thus be tiled by dominoes.

For knights on a 7×7 chessboard, notice that in each move, a knight lands on a square of the opposite color from where it started, so simultaneous legal moves would be possible only for a board with the same number of black and white squares. But a 7×7 chessboard does not have an equal number of black and white squares.

For the Tetris question, remember that the seven Tetris quadrominoes correspond to the shapes of the letters *O, I, L, J, T, S,* and *Z*. Notice that all the Tetris pieces cover the same number of black and white squares except for one—the T-shaped piece—so covering a 4×7 chessboard with them is impossible.

For the 8×8×8 cube with opposite corners removed, using coordinates to specify the location of the 1×1×1 cubes, color the 1×1×1 cube at position (i, j, k) with one of three colors, corresponding to the remainder of $i + j + k$ when you divide it by 3. Since a 1×1×3 block will then cover exactly one 1×1×1 cube of each color, the figure can be tiled by 1×1×3 blocks only if there are the same number of cubes of each color in the figure. But there are not.

Matsumoto Sliding Block Puzzle. If the tiles in the starting configuration are numbered as 2 for the young woman; 1, 3, 4, and 6 for the vertical dominoes (numbered left to right and top to bottom); 5 for the horizontal domino; and 7, 8, 9, and 10 for the small squares (numbered left to right and top to bottom), then a solution from the starting configuration shown goes like this: 6, 10, 8, 5, 6, 10 (halfway), 8, 6, 5, 7 (up, left), 9, 6, 10 (left, down), 5, 9, 7, 4, 6, 10, 8, 5, 7 (down, right), 6, 4, 1, 2, 3, 9, 7, 6, 3, 2, 1, 4, 8, 10 (right, up), 5, 3, 6, 8, 2, 9, 7 (up, left), 8, 6, 3, 10 (right, down), 2, 9 (down, right), 1, 4, 2, 9, 7 (halfway), 8, 6, 3, 10, 9 (down), 2, 4, 1, 8, 7, 6, 3, 2, 7, 8, 1, 4, 7 (left, up), 5, 9, 10, 2, 8, 7, 5, 10 (up, left), 2.

Shoelace Clock. 1. There is a method for measuring 3.75 minutes. Suppose the shoelace has endpoints *A* and *B*. Since it is symmetrical, a cut at its midpoint will produce two identical laces that are each possibly nonsymmetrical and have a burn time of 30 minutes. Lay them side by side such that *A* and *B* are side by side. Burn both ends of one lace. After 15 minutes, where they meet and burn out, cut the other lace at the corresponding point. Now you have two laces with no relationship with each other except that they both have a burn time of 15 minutes. Simultaneously light both ends of one lace and one end of the other. When the flames on the first lace meet, after 7.5 minutes, extinguish the flame on the other lace. What remains is a lace that will burn in 7.5 minutes. Lighting both ends will measure 3.75 minutes, when the flames meet.

2. You can measure arbitrarily short intervals of time! For instance, you can measure any time interval that is 60 minutes divided by a power of 2. To see this, cut the identical, symmetric shoelaces at their midpoints to produce four identical nonsymmetric shoelaces. Ignore one, and you have three laces.

We will outline a procedure that will take three laces, two identical, of equal burn time *T* and produce a set of three laces, two identical, of burn time *T*/2. Call the laces lace 1, lace 2, and lace 3, each with burn time *T*, where lace 1 and lace 2 are identical and laid side by side in that manner. Simultaneously light both ends of lace 3 and one end of lace 2. When the flames on lace 3 meet and burn out, extinguish the flame on lace 2 and cut lace 1 at the corresponding position. You now have three laces with burn time $T/2$, two of which are identical.

You can continue this process indefinitely to produce laces with a burn time of the form $T/2^k$.

Vickrey Auction. Here's why a bidder's best strategy is to bid V, what she thinks the car is worth. Let M be the (unknown) maximum of all other bids. No matter what M is, we'll show that bidding any other amount B is never better than bidding V. If both V and B are less than M, the bidder loses the car in either instance, and if both V and B are greater than M, the bidder wins the car and pays M in either instance. So the only scenario where there's a difference in outcome between bidding B and bidding V is if M lies between B and V.

If $B > M > V$, then bidding B is bad for the bidder because she will win the auction but will pay M, more than she feels the car is worth, hence losing money on that transaction; whereas if she had bid V, she would have lost the auction and had no net change to her fortune.

If $B < M < V$, then bidding B is bad for the bidder because she'll lose the auction and the net change to her fortune will be zero, whereas if she had bid her true value V, she would have won the auction and paid M, less than she thought the car was worth, netting a gain to her fortune.

Pentomino Sudoku. See below.

5	3	4	1	2	5	1	2	4	3
1	4	5	3	4	2	5	3	2	1
1	2	3	2	3	5	1	5	4	4
3	5	4	5	2	1	4	3	1	2
4	2	1	3	1	3	2	5	5	4
2	5	2	4	5	1	4	1	3	3
3	1	5	2	3	4	5	4	1	2
4	1	3	5	1	3	2	4	2	5
5	4	1	4	5	2	3	2	3	1
2	3	2	1	4	4	3	1	5	5

Power Indices. The six orderings of three groups are *ABC, ACB, BAC, BCA, CAB, CBA*. If group *A* is size 48, *B* is size 49, and *C* is size 3, then the middle group in each ordering is pivotal. So the power of each group, according to the Shapley-Shubik index, is 1/3.

Unknown Polynomial. You need only two questions to determine the polynomial. First ask for the value of the polynomial at 1. The answer will give you the sum of the coefficients, which must be larger than every coefficient if they are all nonnegative. If the number of digits in the answer is k, then ask for the value of the polynomial at 10^{k+1}. The digits of the answer will display the coefficients of each term within blocks of size $k + 1$. For example, if the value of the polynomial at 1 is 1044, then the largest coefficient is not more than four digits long. Now ask for the value of the polynomial at 10^5. If the answer is 1200345006780009, then counting from the right, each block of size 5 displays a coefficient of the polynomial, which must therefore be $12x^3 + 345x^2 + 678x + 9$.

Five Points on a Sphere. Choose any pair of points from our collection of five points. These points determine a great circle that divides the sphere into two hemispheres (a great circle on the sphere is a circle whose center is the center of the sphere). The two points thus lie on the boundary of both hemispheres. Of the three other points in our collection, at least two must be contained in one of these hemispheres. So that hemisphere contains those two points plus both of the initial pair of points—all together, that's four points.

notes

CHAPTER 1. FLOURISHING

Epigraph. Simone Weil, *Gravity and Grace,* trans. A. Wills (New York: G. P. Putnam's Sons, 1952), 188.

1. From a letter Simone wrote to Father Perrin, which can be found in a collection of her writings: *Waiting for God,* trans. Emma Craufurd (London: Routledge & K. Paul, 1951), 64.
2. An excellent summary of Simone's connection between mathematics and her spirituality can be found in Scott Taylor, "Mathematics and the Love of God: An Introduction to the Thought of Simone Weil," available at http://colby.edu/~sataylor/SimoneWeil.pdf.
3. See Maurice Mashaal, *Bourbaki: A Secret Society of Mathematicians* (Providence: American Mathematical Society, 2006), 109–13.
4. Simone and André Weil's relationship is explored in the memoir by Sylvie Weil, André's daughter: *At Home with André and Simone Weil,* trans. Benjamin Ivry (Evanston, IL: Northwestern University Press, 2010).
5. Throughout 2018, the top four public companies by market capitalization were technology companies: Apple, Alphabet (the parent company of Google), Microsoft, and Amazon. In addition, the top ten included three more: Tencent, Alibaba, and Facebook.
6. Michael Barany, "Mathematicians Are Overselling the Idea That 'Math Is Everywhere,'" *Guest Blog, Scientific American,* August 16, 2016, https://blogs.scientificamerican.com/guest-blog/mathematicians-are-overselling-the-idea-that-math-is-everywhere/.
7. See, for instance, Andrew Hacker, "Is Algebra Necessary?," editorial, *New York Times,* July 28, 2012, https://www.nytimes.com/2012/07/29/opinion/sunday/is-algebra-necessary.html; E. O. Wilson, "Great Scientist ≠ Good at Math," editorial, *Wall Street Journal,* April 5, 2013, https://www.wsj.com

/articles/SB10001424127887323611604578398943650327184. Both of these rely on flawed understandings of what mathematics really is.

8. Two of the more recent examples are the reports *A Common Vision for Undergraduate Mathematical Sciences Programs in 2025* (2015), published by the Mathematical Association of America and available at https://www.maa.org/sites/default/files/pdf/CommonVisionFinal.pdf, and *Catalyzing Change in High School Mathematics: Initiating Critical Conversations* (2018), by the National Council of Teachers of Mathematics, available for purchase at https://www.nctm.org/catalyzing/.
9. See Christopher J. Phillips, *The New Math: A Political History* (Chicago: University of Chicago Press, 2015).
10. See, for example, Robert P. Moses and Charles E. Cobb Jr., *Radical Equations: Civil Rights from Mississippi to the Algebra Project* (Boston: Beacon, 2002), ch. 1.
11. For a sobering assessment, read Cathy O'Neil, *Weapons of Math Destruction: How Big Data Increases Inequality and Threatens Democracy* (New York: Crown, 2016).
12. Erin A. Maloney, Gerardo Ramirez, Elizabeth A. Gunderson, Susan C. Levine, and Sian L. Beilock, "Intergenerational Effects of Parents' Math Anxiety on Children's Math Achievement and Anxiety," *Psychological Science* 26, no. 9 (2015): 1480–88.
13. "Definitions," trans. D. S. Hutchinson, in Plato, *Complete Works,* ed. John M. Cooper (Indianapolis: Hackett, 1997), 1680.
14. Some notable examples include Ubiratan D'Ambrosio, who emphasized the sociocultural dimensions of math education and initiated the study of ethno-mathematics in "Socio-cultural Bases for Mathematical Education," in *Proceedings of the Fifth International Congress on Mathematical Education,* ed. Marjorie Carss (Boston: Birkhäuser, 1986), 1–6; Reuben Hersh, who expounds a humanist philosophy of math in *What Is Mathematics, Really?* (Oxford: Oxford University Press, 1997); and Rochelle Gutiérrez, who addresses structures, policies, and practices that have dehumanized mathematics education, especially for students of color, in "The Need to Rehumanize Mathematics," in *Rehumanizing Mathematics for Black, Indigenous, and Latinx Students: Annual Perspectives in Mathematics Education,* ed. Imani Goffney and Gutiérrez (Reston, VA: National Council of Teachers of Mathematics, 2018), 1–10.
15. Joshua Wilkerson, "Cultivating Mathematical Affections: Developing a

Productive Disposition through Engagement in Service-Learning" (PhD thesis, Texas State University, 2017), 1, https://digital.library.txstate.edu/handle/10877/6611.

CHAPTER 2. EXPLORATION

Epigraph 1. Maryam Mirzakhani, quoted in Bjorn Carey, "Stanford's Maryam Mirzakhani Wins Fields Medal," *Stanford News,* August 12, 2014, https://news.stanford.edu/news/2014/august/fields-medal-mirzakhani-081214.html.

Epigraph 2. Eugenia Cheng, *How to Bake Pi* (New York: Basic Books, 2015), 2.

1. John Joseph Fahie, *Galileo: His Life and Work* (New York: James Pott, 1903), 114.
2. See, for instance, Blaine Friedlander, "To Keep Saturn's A Ring Contained, Its Moons Stand United," *Cornell Chronicle,* October 16, 2017, http://news.cornell.edu/stories/2017/10/keep-saturns-ring-contained-its-moons-stand-united, and "Giant Planets in the Solar System and Beyond: Resonances and Rings" (Cornell Astronomy Summer REU Program, 2012), http://hosting.astro.cornell.edu/specialprograms/reu2012/workshops/rings/.
3. A more well-developed example appears in Paul Lockhart's "A Mathematician's Lament" (2002), available on the blog *Devlin's Angle:* Keith Devlin, "Lockhart's Lament," March 2008, https://www.maa.org/external_archive/devlin/devlin_03_08.html.
4. Achi and other games from Africa can be found in physical form in the MIND Research Institute's South of the Sahara game box, available at https://www.mindresearch.org/mathminds-games.
5. I have not seen any definitive source that resolves these ambiguities.
6. See Claudia Zaslavsky, *Math Games & Activities from Around the World* (Chicago: Chicago Review Press, 1998).
7. Fawn Nguyen, "These Twenty Things," *Finding Ways* (blog), December 19, 2016, http://fawnnguyen.com/these-twenty-things/.
8. See, for instance, Kevin Hartnett, "Mathematicians Seal Back Door to Breaking RSA Encryption," *Abstractions Blog, Quanta Magazine,* December 17, 2018, https://www.quantamagazine.org/mathematicians-seal-back-door-to-breaking-rsa-encryption-20181217/; Rama Mishra and Shantha Bhushan, "Knot Theory in Understanding Proteins," *Journal of Mathematical Biology* 65, nos. 6–7 (December 2012): 1187–213, available at https://link

.springer.com/article/10.1007/s00285-011-0488-3; Chris Budd and Cathryn Mitchell, "Saving Lives: The Mathematics of Tomography," *Plus Magazine,* June 1, 2008, https://plus.maths.org/content/saving-lives-mathematics-tomography.

9. The *Art of Problem Solving* website is a good resource: https://artofproblemsolving.com/.
10. Ben Orlin, *Math with Bad Drawings* (New York: Black Dog & Leventhal, 2018), 10–12.
11. You may remember wayfinding being featured in Disney's movie *Moana* (2016). "Star of gladness" refers to the star Arcturus.
12. See Richard Schiffman, "Fantastic Voyage: Polynesian Seafaring Canoe Completes Its Globe-Circling Journey," *Scientific American,* June 13, 2017, https://www.scientificamerican.com/article/fantastic-voyage-polynesian-seafaring-canoe-completes-its-globe-circling-journey/.
13. Linda updated this quote from her interview with Cheryl Ernst, "Ethnomathematics Makes Difficult Subject Relevant," *Mālamalama,* July 15, 2010, http://www.hawaii.edu/malamalama/2010/07/ethnomathematics/.

CHAPTER 3. MEANING

Epigraph 1. *Sónya Kovalévsky: Her Recollections of Childhood,* trans. Isabel F. Hapgood (New York: Century, 1895), 316.

Epigraph 2. Jorge Luis Borges, *This Craft of Verse* (Cambridge, MA: Harvard University Press, 2002), 22.

1. For an amusing video of a similar situation, do an online search for news coverage of President Obama's car getting stuck as it rolled over an exit ramp at the US Embassy in Dublin in May 2011.
2. Henri Poincaré, *Science and Hypothesis,* trans. William John Greenstreet (New York: Walter Scott, 1905), 141.
3. Jo Boaler, "Memorizers Are the Lowest Achievers and Other Common Core Math Surprises," editorial, *Hechinger Report,* May 7, 2015, https://hechingerreport.org/memorizers-are-the-lowest-achievers-and-other-common-core-math-surprises/.
4. Robert P. Moses and Charles E. Cobb Jr., *Radical Equations: Civil Rights from Mississippi to the Algebra Project* (Boston: Beacon, 2002), 119–22.
5. See Cassius Jackson Keyser, *Mathematics as a Culture Clue, and Other Essays* (New York: Scripta Mathematica, Yeshiva University, 1947), 218.

6. William Byers, *How Mathematicians Think: Using Ambiguity, Contradiction, and Paradox to Create Mathematics* (Princeton: Princeton University Press, 2007).
7. This definition was popularized by the mathematician Keith Devlin in *Mathematics: The Science of Patterns* (New York: Scientific American Library, 1997) and may have originated in Lynn Steen, "The Science of Patterns," *Science* 240, no. 4852 (April 29, 1988): 611–16.

CHAPTER 4. PLAY

Epigraph 1. Martin Buber, *Pointing the Way: Collected Essays,* ed. and trans. Maurice S. Friedman (New York: Harper & Row, 1963), 21.

Epigraph 2. Attributed to Sophie Germain by Count Guglielmo Libri-Carducci in his eulogy for her. See Ioan James, *Remarkable Mathematicians: From Euler to Von Neumann* (New York: Cambridge University Press, 2002), 58.

1. Johan Huizinga, *Homo Ludens: A Study in the Play-Element of Culture,* translated from the German [translator unknown] (London: Routledge & Kegan Paul, 1949).
2. G. K. Chesterton, *All Things Considered* (London: Methuen, 1908), 96.
3. Huizinga, *Homo Ludens,* 8.
4. Paul Lockhart, "A Mathematician's Lament" (2002), 4, available at *Devlin's Angle:* Keith Devlin, "Lockhart's Lament," March 2008, https://www.maa.org/external_archive/devlin/devlin_03_08.html.
5. For a description of the modeling cycle, see, for instance, *GAIMME: Guidelines for Assessment and Instruction in Mathematical Modeling Education,* 2nd ed., ed. Sol Garfunkel and Michelle Montgomery, Consortium for Mathematics and Its Applications and the Society for Industrial and Applied Mathematics (Philadelphia, 2019), available at https://www.siam.org/Publications/Reports/Detail/guidelines-for-assessment-and-instruction-in-mathematical-modeling-education.
6. Blaise Pascal, *Pensées,* trans. W. F. Trotter (New York: E. P. Dutton, 1958), 4, no. 10.
7. To answer the question, it will be helpful to first convince yourself that only the last two digits of both numbers affect the last two digits of the product (think about how multiplication is done). So it's enough to check the square of an ending to see if it is stubborn. To check if 21 is stubborn, square 21 and see if its last digits are 21. They are not. Then it's helpful to

realize that you don't need to test all hundred possibilities for two-digit endings. That's because a stubborn two-digit ending must have a stubborn one-digit ending. But there are only four such endings: 0, 1, 5, 6. So you need to check only the two-digit endings that end in 0, 1, 5, 6.

8. It may surprise you that among all 10^{15} possible fifteen-digit endings, there are only four that are stubborn! Here they are:

   . . . 000000000000000,
   . . . 000000000000001,
   . . . 259918212890625,
   . . . 740081787109376.

   Now, what do you notice about them? What do you wonder? Are there patterns?

9. These numbers are known in the mathematical literature as *automorphic numbers.* They are also related to p-adic numbers when the base is prime.
10. Simone Weil, *Waiting for God,* trans. Emma Craufurd (London: Routledge & K. Paul, 1951), 106.
11. G. H. Hardy, *A Mathematician's Apology* (Cambridge: Cambridge University Press, 1940).
12. See the prime minister's remarks at Sarah Polus, "Full Transcript: Prime Minister Lee Hsien Loong's Toast at the Singapore State Dinner," *Washington Post,* August 2, 2016, https://www.washingtonpost.com/news/reliable-source/wp/2016/08/02/full-transcript-prime-minister-lee-hsien-loongs-toast-at-the-singapore-state-dinner/.
13. "Republic," trans. Paul Shorey, in *The Collected Dialogues of Plato,* ed. Edith Hamilton and Huntington Cairns (Princeton: Princeton University Press, 1961), 768 (7.536e).

CHAPTER 5. BEAUTY

Epigraph 1. "Autobiography of Olga Taussky-Todd," ed. Mary Terrall (Pasadena, California, 1980), Oral History Project, California Institute of Technology Archives, 6; available at http://resolver.caltech.edu/CaltechOH:OH_Todd_O.

Epigraph 2. Quoted in Donald J. Albers, "David Blackwell," in *Mathematical People: Profiles and Interviews,* ed. Albers and Gerald L. Alexanderson (Wellesley, MA: A. K. Peters, 2008), 21.

1. "Interview with Research Fellow Maryam Mirzakhani," *Clay Mathematics Institute Annual Report 2008,* https://www.claymath.org/library/annual_report/ar2008/08Interview.pdf, 13.
2. Semir Zeki, John Paul Romaya, Dionigi M. T. Benincasa, and Michael F. Atiyah, "The Experience of Mathematical Beauty and Its Neural Correlates," *Frontiers in Human Neuroscience* 8 (2014): 68.
3. G. H. Hardy, *A Mathematician's Apology* (Cambridge: Cambridge University Press, 1940); Harold Osborne, "Mathematical Beauty and Physical Science," *British Journal of Aesthetics* 24, no. 4 (Autumn 1984): 291–300; William Byers, *How Mathematicians Think: Using Ambiguity, Contradiction, and Paradox to Create Mathematics* (Princeton: Princeton University Press, 2007).
4. Paul Erdős, quoted in Paul Hoffman, *The Man Who Loved Only Numbers: The Story of Paul Erdős and the Search for Mathematical Truth* (London: Fourth Estate, 1998), 44.
5. Martin Gardner, "The Remarkable Lore of the Prime Numbers," Mathematical Games, *Scientific American* 210 (March 1964): 120–28.
6. Erdős reportedly quipped, "You don't have to believe in God, but you should believe in The Book" (Hoffman, *Man Who Loved Only Numbers,* 26). In homage to Erdős, Martin Aigner and Günter Ziegler's collection of elegant proofs of various theorems is playfully titled *Proofs from THE BOOK* (New York: Springer, 2010).
7. Sydney Opera House Trust, "The Spherical Solution," https://www.sydneyoperahouse.com/our-story/sydney-opera-house-history/spherical-solution.html.
8. Jordan Ellenberg, *How Not to Be Wrong: The Power of Mathematical Thinking* (New York: Penguin, 2014), 436–37.
9. Albert Einstein, *Ideas and Opinions* (New York: Crown, 1954), 233.
10. Erica Klarreich, "Mathematicians Chase Moonshine's Shadow," *Quanta Magazine,* March 12, 2015, https://www.quantamagazine.org/mathematicians-chase-moonshine-string-theory-connections-20150312/.
11. Simon Singh, "Interview with Richard Borcherds," *The Guardian,* August 28, 1998, https://simonsingh.net/media/articles/maths-and-science/interview-with-richard-borcherds/.
12. C. S. Lewis, *The Weight of Glory* (New York: Macmillan, 1949), 7.
13. Barbara Oakley, "Make Your Daughter Practice Math. She'll Thank You

Later," editorial, *New York Times,* August 7, 2018, https://www.nytimes.com/2018/08/07/opinion/stem-girls-math-practice.html.

CHAPTER 6. PERMANENCE

Epigraph 1. Bernhard Riemann, "On the Psychology of Metaphysics: Being the Philosophical Fragments of Bernhard Riemann," trans. C. J. Keyser, *The Monist* 10, no. 2 (1900): 198.

Epigraph 2. Network of Minorities in Mathematical Sciences, "Tai-Danae Bradley: Graduate Student, CUNY Graduate Center," *Mathematically Gifted and Black,* http://mathematicallygiftedandblack.com/rising-stars/tai-danae-bradley/.

1. The word *law* is sometimes also used in reference to mathematical ideas—usually either empirically observed patterns that have been validated by a theorem (e.g., the law of large numbers) or axioms that are being assumed as a foundation for knowledge (e.g., the commutative law, the law of the excluded middle).
2. David Eugene Smith, "Religio Mathematici," *American Mathematical Monthly* 28, no. 10 (1921): 341.
3. Morris Kline, *Mathematics for the Nonmathematician* (New York: Dover, 1985), 9.
4. Read more about *The Art of Gaman,* curated by Delphine Hirasuna, in Susan Stamberg, "The Creative Art of Coping in Japanese Internment," NPR, May 12, 2010, https://www.npr.org/templates/story/story.php?storyId=126557553.
5. An eighty-one-step solution, starting from a slightly different starting configuration, was printed in Martin Gardner, "The Hypnotic Fascination of Sliding Block Puzzles," Mathematical Games, *Scientific American* 210 (February 1964): 122–30. A solution starting from Matsumoto's starting configuration can be found in this book's Hints & Solutions to Puzzles.
6. George Orwell, *1984* (Boston: Houghton Mifflin Harcourt, 1949), 76.

CHAPTER 7. TRUTH

Epigraph 1. John 18:38 (Good News Translation).

Epigraph 2. Blaise Pascal, *Pensées,* trans. W. F. Trotter (New York: E. P. Dutton, 1958), 259, no. 864.

1. Hannah Arendt, "Truth and Politics," *New Yorker,* February 25, 1967, reprinted in Arendt, *Between Past and Future* (New York: Penguin, 1968), 257.
2. See Michael P. Lynch, *True to Life: Why Truth Matters* (Cambridge, MA: MIT Press, 2004).
3. Errors sometimes lead to further exploration. The correct calculation is $777 \times 1{,}144 = 888{,}888$. Interesting! But the mistyped calculation $777 \times 144 = 111{,}888$ also has a noteworthy pattern. What's going on here?
4. Gian-Carlo Rota, "The Concept of Mathematical Truth," *Review of Metaphysics* 44, no. 3 (March 1991): 486.
5. Eugene Wigner, "The Unreasonable Effectiveness of Mathematics in the Natural Sciences," *Communications on Pure and Applied Mathematics* 13 (1960): 14.
6. Kenneth Burke, "Literature as Equipment for Living," collected in *The Philosophy of Literary Form: Studies in Symbolic Action* (Baton Rouge: Louisiana State University Press, 1941), 293–304.
7. Quoted in David Brewster, *The Life of Sir Isaac Newton* (New York: J. & J. Harper, 1832), 300–301.

CHAPTER 8. STRUGGLE

Epigraph 1. Simone Weil, *Waiting for God,* trans. Emma Craufurd (London: Routledge & K. Paul, 1951), 107.

Epigraph 2. Martha Graham, "An Athlete of God," in *This I Believe: The Personal Philosophies of Remarkable Men and Women,* ed. Jay Allison and Dan Gediman, with John Gregory and Viki Merrick (New York: Holt, 2006), 84.

1. Alasdair MacIntyre, *After Virtue: A Study in Moral Theory,* 3rd ed. (South Bend, IN: University of Notre Dame Press, 2007), 188.
2. Ibid.
3. See Eric M. Anderman, "Students Cheat for Good Grades. Why Not Make the Classroom about Learning and Not Testing?," *The Conversation,* May 20, 2015, https://theconversation.com/students-cheat-for-good-grades-why-not-make-the-classroom-about-learning-and-not-testing-39556.
4. Carol Dweck, "The Secret to Raising Smart Kids," *Scientific American,* January 1, 2015, https://www.scientificamerican.com/article/the-secret-to-raising-smart-kids1/.
5. Teachers will find an excellent resource on how mindsets affect learning in

mathematics, as well as practical suggestions for how to change mindsets, in Jo Boaler, *Mathematical Mindsets* (San Francisco: Jossey-Bass, 2016).

6. "Interview with Maryam Mirzakhani," *Clay Math Institute Annual Report 2008*, https://www.claymath.org/library/annual_report/ar2008/08Interview.pdf.
7. David Richeson, "A Conversation with Timothy Gowers," *Math Horizons* 23, no. 1 (September 2015): 10–11.
8. Laurent Schwartz, *A Mathematician Grappling with His Century* (Basel: Birkhauser, 2001), 30.

CHAPTER 9. POWER

Epigraph 1. Quoted in Stephen Winsten, *Days with Bernard Shaw* (New York: Vanguard, 1949), 291.

Epigraph 2. Augustus de Morgan, quoted in Robert Perceval Graves, *The Life of Sir William Rowan Hamilton*, vol. 3 (Dublin: Dublin University Press, 1889), 219.

1. See Isidor Wallimann, Howard Rosenbaum, Nicholas Tatsis, and George Zito, "Misreading Weber: The Concept of 'Macht,'" *Sociology* 14, no. 2 (May 1980): 261–75.
2. Andy Crouch, *Playing God: Redeeming the Gift of Power* (Downers Grove, IL: InterVarsity Press, 2014), 17.
3. Thanks to my friend Lew Ludwig for pointing this out to me.
4. Dave Bayer and Persi Diaconis, "Trailing the Dovetail Shuffle to Its Lair," *Annals of Applied Probability* 2, no. 2 (May 1992): 294–313.
5. These details can be found in Karen D. Rappaport, "S. Kovalevsky: A Mathematical Lesson," *American Mathematical Monthly* 88, no. 8 (October 1981): 564–74.
6. Erica N. Walker, *Beyond Banneker: Black Mathematicians and the Paths to Excellence* (Albany: SUNY Press, 2014).
7. Cathy O'Neil, *Weapons of Math Destruction: How Big Data Increases Inequality and Threatens Democracy* (New York: Crown, 2016).
8. Parker J. Palmer, *The Courage to Teach: Exploring the Inner Landscape of a Teacher's Life*, 10th anniversary ed. (San Francisco: Jossey-Bass, 2007), 7.

CHAPTER 10. JUSTICE

Epigraph. Simone Weil, *Gravity and Grace*, trans. A. Wills (New York: G. P. Putnam's Sons, 1952), 188.

1. E.g., Timothy Keller, *Generous Justice: How God's Grace Makes Us Just* (New York: Penguin, 2012).
2. Take the tests at https://implicit.harvard.edu/implicit/.
3. Victor Lavy and Edith Sands, "On the Origins of Gender Gaps in Human Capital: Short- and Long-Term Consequences of Teachers' Biases," *Journal of Public Economics* 167 (2018): 263–79.
4. Michela Carlana, "Implicit Stereotypes: Evidence from Teachers' Gender Bias," *Quarterly Journal of Economics* (forthcoming): https://doi.org/10.1093/qje/qjz008.
5. In 2004, about one-third of students who entered US universities intended to major in a STEM field. Of those, the six-year completion rate was about 45 percent for white and Asian students, and 25 percent for others. There's a lot of interesting data in Kevin Eagan, Sylvia Hurtado, Tanya Figueroa, and Bryce Hughes, "Examining STEM Pathways among Students Who Begin College at Four-Year Institutions," paper commissioned for the Committee on Barriers and Opportunities in Completing 2-Year and 4-Year STEM Degrees (Washington DC: National Academies Press, 2014), https://sites.nationalacademies.org/cs/groups/dbassesite/documents/webpage/dbasse_088834.pdf.
6. See Jennifer Engle and Vincent Tinto, *Moving beyond Access: College Success for Low-Income, First-Generation Students* (Washington DC: Pell Institute, 2008), https://files.eric.ed.gov/fulltext/ED504448.pdf.
7. For instance, among US citizens who earned math PhDs in 2015, 84 percent were white and 72 percent were men. See William Yslas Vélez, Thomas H. Barr, and Colleen A. Rose, "Report on the 2014–2015 New Doctoral Recipients," *Notices of the AMS* 63, no. 7 (August 2016): 754–65.
8. See "Finally, an Asian Guy Who's Good at Math (Part Two)," *Angry Asian Man* (blog), January 4, 2016, http://blog.angryasianman.com/2016/01/finally-asian-guy-whos-good-at-math.html.
9. Rochelle Gutiérrez, "Enabling the Practice of Mathematics Teachers in Context: Toward a New Equity Research Agenda," *Mathematical Thinking and Learning* 4, nos. 2–3 (2002): 147.
10. See, for instance, National Council of Teachers of Mathematics, *Catalyz-*

ing Change in High School Mathematics: Initiating Critical Conversations (Reston, VA : The National Council of Teachers of Mathematics, 2018); Jo Boaler, "Changing Students' Lives through the De-tracking of Urban Mathematics Classrooms," *Journal of Urban Mathematics Education* 4, no. 1 (July 2011): 7–14.

11. William F. Tate, "Race, Retrenchment, and the Reform of School Mathematics," *Phi Delta Kappan* 75, no. 6 (February 1994): 477–84.

CHAPTER 11. FREEDOM

Epigraph 1. Helen Keller, *The Story of My Life* (New York: Grosset & Dunlap, 1905), 39.

Epigraph 2. Eleanor Roosevelt, *You Learn by Living* (New York: Harper & Row, 1960), 152.

1. To learn some of his shortcuts, see Arthur Benjamin and Michael Shermer, *Secrets of Mental Math* (New York: Three Rivers, 2006).
2. This shortcut for multiplying by 11 will require a "carry" if the sum of the digits is 10 or more. For instance, to compute 75×11, you should add 7 and 5 to get 12, put the 2 between the 7 and the 5, and then carry the 1 by adding it to the 7, to get 8. Thus, the answer is 825. If you know some algebra, you can use it to show why the shortcut works: the number $10a + b$ is the number with digits a and b. Then $(10a + b) \times 11 = 110a + 11b = 100a + 10(a + b) + b$. This last expression does indeed suggest adding the two digits and putting their sum between them.
3. Georg Cantor, "Foundations of a General Theory of Manifolds: A Mathematico-Philosophical Investigation into the Theory of the Infinite," trans. William Ewald, in *From Kant to Hilbert: A Source Book in the Foundations of Mathematics*, ed. Ewald (New York: Oxford University Press, 1996), vol. 2, 896 (§8). Italics in the original.
4. Evelyn Lamb, "A Few of My Favorite Spaces: The Infinite Earring," *Roots of Unity* (blog), *Scientific American*, July 31, 2015, https://blogs.scientificamerican.com/roots-of-unity/a-few-of-my-favorite-spaces-the-infinite-earring/.
5. J. W. Alexander, "An Example of a Simply Connected Surface Bounding a Region Which Is Not Simply Connected," *Proceedings of the National Academy of Sciences of the United States of America* 10, no. 1 (January 1924): 8–10.

6. Robert Rosenthal and Lenore Jacobson, "Teachers' Expectancies: Determinants of Pupils' IQ Gains," *Psychological Reports* 19 (1966): 115–18. It is worth noting that this study has attracted controversy. An interesting account of this study, including critiques and follow-up studies, can be found in Katherine Ellison, "Being Honest about the Pygmalion Effect," *Discover Magazine,* October 29, 2015, http://discovermagazine.com/2015/dec/14-great-expectations.
7. bell hooks, *Teaching to Transgress: Education as the Practice of Freedom* (New York: Routledge, 1994), 3.
8. Ibid.

CHAPTER 12. COMMUNITY

Epigraph 1. Bill Thurston, October 30, 2010, reply to "What's a Mathematician To Do?," *Math Overflow,* https://mathoverflow.net/questions/43690/whats-a-mathematician-to-do.

Epigraph 2. Deanna Haunsperger, "The Inclusion Principle: The Importance of Community in Mathematics," MAA Retiring Presidential Address, Joint Mathematics Meeting, Baltimore, January 19, 2019; video available at https://www.youtube.com/watch?v=jwAE3iHi4vM.

1. Parker Palmer, *To Know as We Are Known* (New York: Harper Collins, 1993), 9.
2. See Gina Kolata, "Scientist at Work: Andrew Wiles; Math Whiz Who Battled 350-Year-Old Problem," *New York Times,* June 29, 1993, https://www.nytimes.com/1993/06/29/science/scientist-at-work-andrew-wiles-math-whiz-who-battled-350-year-old-problem.html. Wiles's error was fixed a couple of years later with the help of Richard Taylor.
3. See Dennis Overbye, "Elusive Proof, Elusive Prover: A New Mathematical Mystery," *New York Times,* August 15, 2006, https://www.nytimes.com/2006/08/15/science/15math.html.
4. See Thomas Lin, "After Prime Proof, an Unlikely Star Rises," *Quanta Magazine,* April 2, 2015, https://www.quantamagazine.org/yitang-zhang-and-the-mystery-of-numbers-20150402/.
5. Jerrold W. Grossman, "Patterns of Collaboration in Mathematical Research," *SIAM News* 35, no. 9 (November 2002): 8–9; also available at https://archive.siam.org/pdf/news/485.pdf.
6. I'll mention just a few programs here that may appeal widely. There are

more than two hundred "math circles" throughout the US that gather children periodically for discovery and excitement around low-threshold, high-ceiling problems and interactive exploration; you can find a group on the National Association of Math Circles website (http://www.mathcircles.org/). BEAM (Bridge to Enter Advanced Mathematics; https://www.beammath.org/) offers day and residential programs designed to help underserved students enter the scientific professions. In the past I've taught at a math camp called MathPath (http://www.mathpath.org/), which brings middle school kids together each summer for a mix of math and outdoor activities; programs like this exist at all educational levels. The Park City Mathematics Institute (https://www.ias.edu/pcmi) has a three-week summer program for math teachers (and other groups in the math community) to reflect on math teaching and leadership.

7. See, for instance, Talithia Williams, *Power in Numbers: The Rebel Women of Mathematics* (New York: Race Point, 2018); *101 Careers in Mathematics*, ed. Andrew Sterrett, 3rd ed. (Washington DC: Mathematical Association of America, 2014).
8. Simone Weil, letter to Father Perrin, collected in *Waiting for God*, trans. Emma Craufurd (London: Routledge & K. Paul, 1951), 64.
9. See the *MAA Instructional Practices Guide* (2017), including references, from the Mathematical Association of America, available at https://www.maa.org/programs-and-communities/curriculum%20resources/instructional-practices-guide.
10. See Darryl Yong, "Active Learning 2.0: Making It Inclusive," *Adventures in Teaching* (blog), August 30, 2017, https://profteacher.com/2017/08/30/active-learning-2-0-making-it-inclusive/.
11. Ilana Seidel Horn's book *Motivated: Designing Math Classrooms Where Students Want to Join In* (Portsmouth, NH: Heinemann, 2017) contains ideas on how to do so.
12. See Justin Wolfers, "When Teamwork Doesn't Work for Women," *New York Times*, January 8, 2016, https://www.nytimes.com/2016/01/10/upshot/when-teamwork-doesnt-work-for-women.html.
13. See Association for Women in Science–Mathematical Association of America Joint Task Force on Prizes and Awards, "Guidelines for MAA Selection Committees: Avoiding Implicit Bias" (prepared August 2011, approved August 2012), Mathematical Association of America, https://www

.maa.org/sites/default/files/pdf/ABOUTMAA/AvoidingImplicitBias_revisionMarch2018.pdf.

14. Karen Uhlenbeck, "Coming to Grips with Success," *Math Horizons* 3, no. 4 (April 1996): 17.

CHAPTER 13. LOVE

Epigraph 1. 1 Corinthians 13:1 (Good News Translation).

Epigraph 2. *The Papers of Martin Luther King, Jr.*, ed. Clayborne Carson, vol. 1, *Called to Serve: January 1929–June 1951*, ed. Ralph E. Lucker and Penny A. Russell (Berkeley: University of California Press, 1992), 124.

1. See, for example, Hannah Fry, *The Mathematics of Love: Patterns, Proofs, and the Search for the Ultimate Equation* (New York: Simon & Schuster, 2015).
2. Simone Weil, letter to Father Perrin, collected in *Waiting for God*, trans. Emma Craufurd (London: Routledge & K. Paul, 1951), 64.
3. I tell this story in Francis Edward Su, "The Lesson of Grace in Teaching," in *The Best Writing on Mathematics 2014*, ed. Mircea Petici (Princeton: Princeton University Press, 2014), 188–97, also available at http://mathyawp.blogspot.com/2013/01/the-lesson-of-grace-in-teaching.html.
4. Simone Weil, "Reflections on the Right Use of School Studies with a View to the Love of God," in *Waiting for God*, trans. Emma Craufurd (London: Routledge & K. Paul, 1951), 115.

ACKNOWLEDGMENTS

Simone Weil's original quote is "L'attention est la forme la plus rare et la plus pure de la générosité." See Weil and Joë Bousquet, *Correspondance* (Lausanne: Editions l'Age d'Homme, 1982), 18.

index